ON EXPERTISE

RSA·STR

THE RSA SERIES IN TRANSDISCIPLINARY RHETORIC

The RSA Series in Transdisciplinary Rhetoric is a collaboration with the Rhetoric Society of America to publish innovative and rigorously argued scholarship on the tremendous disciplinary breadth of rhetoric. Books in the series take a variety of approaches, including theoretical, historical, interpretive, critical, or ethnographic, and examine rhetorical action in a way that appeals, first, to scholars in communication studies and English or writing, and, second, to at least one other discipline or subject area.

Ashley Rose Mehlenbacher

ON EXPERTISE

Cultivating Character, Goodwill, and Practical Wisdom

THE PENNSYLVANIA STATE UNIVERSITY PRESS
UNIVERSITY PARK, PENNSYLVANIA

Library of Congress Cataloging-in-Publication Data

Names: Mehlenbacher, Ashley Rose, 1983– author.
Title: On expertise : cultivating character, goodwill,
    and practical wisdom / Ashley Rose Mehlenbacher.
Other titles: RSA series in transdisciplinary rhetoric.
Description: University Park, Pennsylvania : The Pennsylvania
    State University Press, [2022] | Series: The RSA series in
    transdisciplinary rhetoric | Includes bibliographical
    references and index.
Summary: "A rhetorical account of expertise and expert status
    that is informed by research in allied fields, including
    Science, Technology, and Society studies, psychology,
    education, and philosophy"—Provided by publisher.
Identifiers: LCCN 2021061266 | ISBN 9780271092768
    (hardback) | ISBN 9780271092775 (paperback)
Subjects: LCSH: Expertise. | Rhetoric.
Classification: LCC BF378.E94 M44 2022 | DDC 153.9—
    dc23/eng/20211216
LC record available at https://lccn.loc.gov/2021061266

The Pennsylvania State University Press is a member
of the Association of University Presses.

It is the policy of The Pennsylvania State University Press to
use acid-free paper. Publications on uncoated stock satisfy the
minimum requirements of American National Standard for
Information Sciences—Permanence of Paper for Printed
Library Material, ANSI z39.48–1992.

For Jennifer and Daniel, the teachers; and Daniel Jr., Sebastian Bradley, and Abigail, the students, for all their unfolding and potential expertise. And for Brad, for his goodwill and happiness.

Contents

Contents

Acknowledgments

This book draws on research supported by the Social Sciences and Humanities Research Council of Canada through an Insight Grant; the Ontario Ministry of Research, Innovation and Science through an Early Researcher Award; and the University of Waterloo through an internal seed grant. Related to this research program, the Department of English and Faculty of Arts have provided funding for youth outreach activities. I have appreciated the support of two outstanding department chairs, Kate Lawson and Shelley Hulan, and the department administrative manager, Margaret Ulbrick. I also appreciate course releases I was granted for my participation in developing courses for our Faculty of Science and several engineering programs for the Undergraduate Communication Outcomes Initiative, which allowed me to complete much of the writing for this book. Many research assistants have been involved with this project and, especially, with the book itself, including Ana Patricia Balbon, Danielle Jodway, Devon Moriarty, Paula Núñez de Villavicencio, and Anjiya Sharif. I owe many thanks to the numerous other research assistants who have helped at various stages of this project, including those who have helped me see it through and turn to the next phase in this ongoing research program, now attending specifically to issues in climate change and action, Sara Doody, Sarah Forst, Carolyn Eckert, and Weiran Wang. Crucially, I also want to thank the many participants who spent time completing the survey or interviewing with us to tell us how they understand, develop, enact, and assess expertise in their own lives. In addition to being foundational to the empirical work of this book, participants supplied much insight into the rhetorical complexities of how we approach questions of expertise and expert status. These ideas have generated much thinking on our team about what it is we hope to understand and how best to engage in reflexive methods and approaches to, always imperfectly, gain insight and understanding. Thanks, as well, to Ryan Peterson, at Penn State University Press, and the Rhetoric Society of America Transdisciplinary Rhetoric series editors, Michael Bernard-Donals and Leah Ceccarelli, for their support and development of this project, and to the many people at Penn State University Press who saw this book through production. On the improvement of this manuscript from its original form, I thank the anonymous reviewer as well as Johanna Hartelius—without her perceptive read and incisive advice, this book would be a mere shadow of its current form—and Nicholas Taylor and his colleagues at Grit City Creative LLC for their detailed and thoughtful attention to the text and arguments. I am also grateful to those colleagues at the University of Waterloo and in the field who provided opportunities, resources, or conversations that allowed me to develop my thinking around expertise and multidisciplinary work, including Lamees Al Ethari, Natasha Artemeva, Sune Auken, Fraser Easton, Lai-Tze Fan, Randy Allen Harris, George Lamont, Andrea Jonahs, Carolyn R. Miller, and Katie Plaisance. Deepest appreciation to my good friend Chelsea Ferriday, always a source of motivation, and to Christopher Kampe, for hours of challenging and energizing conversation, as well as my family and, especially, Brad Mehlenbacher, my colleague, collaborator, and partner.

Introduction | Understanding Expertise

We might think of our family doctor as expert, our lawyer as expert, and our plumber as expert. Each of these people we might work with we hope are, indeed, expert. More distantly, scientists and expert political commentators on the nightly news pour into our lives with expert advice or insights. With so many experts one might wonder which expert advice to follow and which experts to trust. In fact, expertise is a topic of great concern, with Tom Nichols proclaiming its demise in his *Death of Expertise*. Nichols (2017, 28) argues that the death of expertise is "like a national bout of ill temper, a childish rejection of authority in all its forms coupled to an insistence that strongly held opinions are indistinguishable from facts." Even in this line of reasoning, as Nichols himself notes, it is not only the public that is culpable for expertise's demise, citing issues in numerous professions. Consider how soon after the United States had elected its forty-fifth president, and soon after his inauguration, scientists were, quite unbelievably, *protesting* in the streets with signs, coupled with an online movement to challenge the new administration's seemingly anti-science—anti-*expert*—efforts. Expertise had, it seemed, never been so relevant. Some researchers are perhaps more cautious in their assessment of symptoms suggesting the ill health of expertise. Grundmann (2018, 373) argues that the so-called death of expertise, and the "so-called populist backlash against science and expertise" fueling reports of expertise's demise, may be "a figment of the imagination, itself in the land of opinion and post-truth." With the COVID-19 pandemic, expertise has been launched even more significantly into public consciousness and discourse. In either case, it is evident the role of experts and expertise has recently generated much consideration.[1]

The concern with the role of experts in modern life, however, is hardly new. Consider Laski's reflection on *The Limitations of the Expert*. Laski explains that the very notion of expert knowledge, the requirement to attain and maintain such knowledge, necessarily precludes the expert from guiding public life. "The expert tends," Laski (1931, 202) argues, "to make his subject the measure of life, instead of making life the measure of his subject." Laski understands the problem of the expert in public life as one concerned with a kind of practical wisdom, writing that the expert "is an invaluable servant and an impossible master. He can explain the consequences of a proposed policy, indicate its wisdom, measure its danger. He can point out possibilities in a proposed line of action. But it is of the essence of *public wisdom* to take the final initiative out of his hands" (204, emphasis mine).

Widening the idea of experts to be inclusive of "professionals," Schön's *The Reflective Practitioner* (1982) charts the postwar assent of specialists with expertise especially in technical and scientific domains. Indeed, Schön cites the American Academy of the Arts and Sciences journal *Daedalus*, in 1963, which proclaimed, "Everywhere in American life, the professions are triumphant" (Lynn 1963, 649). Such professionals had largely built their persuasive power on a techno-scientific rationality that promised objective, measurable, and progressive outcomes. Throughout the 1970s and into the 1980s, however, a range of failures and disasters catalyzed doubt in professionals. Although this doubt in professionals (and experts) began decades ago, the rise of anti-expert discourse appears to be unabated today.

In addition to critique of experts, overwhelming changes to workforces have generated debate about the role of expertise in the twenty-first century. Lynn's (1963, 651) introduction to the issue of *Daedalus* on professions came with a relevant warning: "Because there are simply not enough professionals to go around, the practitioner of today is perforce burdened with too much work." In *The Atlantic*, Jerry Useem (2019) details a trend to view workers as generalists rather than specialists or experts. From the navy's "hybrid sailors" to online retailer Zappos, where employees are hired into "circles" rather than job titles, expertise seems displaced. Useem characterizes this moment as a transformative one[2] where conventional understandings of expert knowledge, once assessed in terms of education and experience, are being adapted in favor of some conception of flexibility or adaptability in thinking. In such a reconfiguration, experts as traditionally conceived are displaced from their once favored status in the workplace. Whether such an adaption is promising is not yet clear, and the change in and of itself does not warn of anti-expert sentiments. What this example illustrates is that the status of experts and the very idea of expertise continue to be debated.[3]

Conversations about experts suggest that important cultural values are being negotiated. And negotiation of these values is accomplished not in the lab but in civic spheres. To understand this moment, we must understand something of the experts' claim to their status. First, an understanding of how experts are conceived is necessary. How expertise is acquired and measured and how publics assess experts and their expertise are also important topics. The rhetorical negotiations surrounding experts and expertise can be illuminated by arguments for and against expert involvement, the boundaries between experts and nonexperts, and matters of trust and values. In these ways, we come to see that expert status is a highly rhetorical activity, at least in the expert-public dynamic. However, the rhetorical activities required to claim expert status in expert communities are perhaps even more significant, if measured by argumentative challenge. In expert-expert negotiations, expert status is indeed accomplished through rhetorical activities.

Although I am arguing that expert status and expertise are both rhetorical in nature, I do not mean to suggest that expertise does not reflect real capacities. We will see that a rhetorical account in fact means that these capacities are quite real and quite critical as a capacity for becoming expert. Expert status, too, is not simply a public relations campaign, although it may be in some cases. Expert status is, rather, often achieved by a complex negotiation with an audience of epistemic claims, a cultivation of skills, and capacities for deliberation and moral judgment, which we understand to be expertise. To this end, the rhetorical conception of expert status and expertise I wish to advance here builds on the work of rhetorical scholars including Danisch (2010), Hartelius (2011), and Majdik and Keith (2011a, 2011b), as well as philosophers examining virtue ethics, and includes concepts of *ethos* (character), *episteme* (knowledge), *techne* (skill or craft), and *phronesis* (practical wisdom). Expertise, further to expert status, is characterized by how such forms of knowledge are invoked in particular situations, comprising intersecting and changing audiences, rhetors, traditions, institutions, objects, and needs, all through forms of practical wisdom.

Understanding experts and expertise asks us to consider what appear to be common characteristics among a variety of knowledgeable people and their various skills. Further, experts and expertise, although most readily identifiable at the level of individual, are concepts that are constituted also by the communities and the spaces where an expert is situated and when expertise takes place. To understand expertise and expert status, a powerful theoretical framework does not require a unifying principle of expertise. Rather, it calls for an approach that allows for the multiple facets of expertise to be examined with respect to particular situations and the faculty, in a rhetorical sense, or capacity to determine the available means with which to respond appropriately. This book is chiefly concerned with expertise in science, the social sciences, technology, health, and medicine, and explores these fields in academic research primarily, but also industry, public or nonprofit sectors, and, critically, citizen science (the practice of everyday people becoming involved in scientific research). Here, expertise is investigated through the lens of rhetorical studies. Rhetorical studies provide versatile theoretical tools to explore many of the dimensions of expertise and the multidisciplinary literature on expertise. Indeed, rhetorical studies can illuminate the key concepts and definitions while also uncovering who might be deemed an expert when different sets of concepts or definitions are used. Further, rhetorical studies can connect these social aspects of becoming expert to the affinities of the mind that underlie our capacities for becoming and being expert.[4]

With this rhetorical vantage, I examine the cultivation of expert ethos and the training and habituating of the mind through practical wisdom, phronesis, as part of the acquisition of expertise and its ongoing enactment. Further, the concept itself reminds us of the socially situated nature of our ethical commitments. Phronesis is a capacity for moral reasoning constrained by the historical, social, and cultural conditions we inhabit, which is to say the account here is descriptive and not prescriptive, as phronesis itself requires ongoing commitment to new and diverse experience to challenge those common beliefs of our time that are unjust. Such an approach has pragmatic as well as theoretical contributions. For example, it can help us understand why some people do not trust experts, what experts can do about that, and how we can negotiate expertise as central to functioning of our expert-reliant lives.

What or Who Is an Expert?

High status occupational titles may also function metonymically for expertise: doctor, lawyer, scientist, and so on. But a profession or specialization is not synonymous with expertise. I would offer the example of a skillful orator as an expert who can help illustrate why equating specialization and expertise unnecessarily conflates two different conceptions. Consider the mastery of the "ability, in each [particular] case, to see the available means of persuasion" (Aristotle, *Rhet.* 1.2.1355b)[5] or the "symbolic means of inducing cooperation in beings that by nature respond to symbols" (Burke 1969, 43) that requires considerable expertise to achieve its most potent effects. Yet rhetoric is not a specialization as we might normally conceive—with a rather more limited or restricted domain of technical knowledge—although a certain technical or theoretical knowledge underlies the field. This is to say, although rhetoricians may be experts in rhetorical craft, they must also become experts[6] in the domain on which they speak in each situation. Here we find an interesting interplay between specialist and technical knowledge. There is the rhetorical craft, a techne and specialist expertise, as well as a generalist expertise of applying this techne for a given situation. Thus, already it is evident that expertise operates in two distinct manners. Although expertise and specialization are not synonymous, they are often concomitant. There is some indication that specialized expert skills are tied to certain domains, and that when these skills are applied outside of said domain they no longer appear to have expert qualities (see Gobet 2016, 238). If a domain is closely related to another, experts still perform better than their nonexpert peers or experts from a domain far afield from the specialty of concern. Thus, the reflective practices that allow experts to excel in their profession are situational and domain-specific (Schön 1983, 167).

Another concept we must distinguish from expertise and expert status is authority. When expertise is called on in a democratic sphere, expertise may seem a cognate of authority.[7] However, expert-to-expert dialectic and deliberation shows a rather different function of expertise. Although expert-to-expert exchanges may still be susceptible to forms of authority, the expert-to-expert deliberation is more immune insofar as the expert's epistemic grounds are nearly equally understood. Miller (2003, 200) explains that Aristotle would not view epistemic negotiations among experts as rhetorical, but rather dialectical, which approximates the kind of rational, *logos*-centric discourse that vernacular accounts of science ascribe to expert intellectual practice. Aristotle, Miller continues, would see such situations as ones where expertise, not ethos, is required: "The intellectual quality needed by the dialectician or the wise person is not *phronesis*, *arete*, or *eunoia*, but *sophia* (wisdom)" (201). Thus, the artistic construction of ethos is unnecessary where there are data, facts, and so on, that might rather be formulated through expert understanding and applied to some problem, the logos-centric advocates would have it. Yet, Miller's analysis of risk assessment as a field and broader studies in rhetoric of science have shown that this is a reductive understanding of how epistemic work is conducted. Miller explains that when the important work classical ethos performs to build trust is not accomplished, trust is lost. In Aristotle's *Rhetoric*, *pistis* provides an important sense of trust, although Garver (2017, 136) reminds us of its rich meaning that can be "rightly translated as proof, argument, reasoning, persuasion, belief, trust, faith, credit, conviction, and confidence." *Pistis* as a state of mind can be achieved through appeals to ethos, *pathos*, and logos (as good reason). In a related field, philosophy of science, Hardwig (1991) describes the importance of trust among researchers and Whyte and Crease (2010, 412) outline trust in expert and nonexpert interactions, defining trust as "deferring with comfort and confidence to others, about something beyond our knowledge or power, in ways that can potentially hurt us." Trust, the rhetoricians and philosophers have noted, is essential to the very enterprise of science itself, and why violations of trust are taken so seriously, including in retractions, for instance.[8] Trust is also necessary for more than expert-public / nonexpert deliberation, and Aristotle might caution us this is why ethical arguments—and not, for instance, inflammatory appeals to the emotions—are an important part of rhetorical thinking. Miller's example also clarifies why expert-public deliberation, too, should not conflate expertise with authority. Speaking from a position of authority for experts is a restricted activity, restricted to their own domain of expertise, and engagement with broader public concerns necessarily means said domain of expertise is not inclusive of the full range of issues. Thus, expert advice[9] is part of deliberative engagement, but so too are the boundaries on expert advice to avoid the encroachment of "techno-scientific rationality" on democratic deliberation.[10]

Recentering the concept of phronesis in the account of ethos provides a critical understanding of how the appeal is more than "mere" persuasion. Rather, ethos is partially an ethical comportment toward one's audience and, thus, insufficient when reconfigured through the logos-centric appeal to expertise that Miller describes. Further to finding phronesis in the *ethotic* appeals, phronesis might also be found in those logos-based configurations of expertise. Although *sophia* or wisdom may initially seem a desirable intellectual virtue for the expert, ultimately it is its more worldly grounded counterpart, phronesis, that haunts the definitions of experts across disciplines. We might locate the "doing of expertise" at the "resolutions of tensions" (as Majdik 2016 argues). When an expert performs to seek such resolution, the expert enacts both individual and collective knowledge. Such enactments are underpinned by configurations of knowledges, inclusive of and, perhaps more provocatively, afforded by phronesis, to resolve our tensions. Moving from the role of phronesis in ethos to the role of phronesis in expertise requires some account of rhetorical theory and virtue ethics, to which we now turn.

What Does Rhetorical Theory Tell Us About Experts and Expertise?

When one imagines an expert, one likely conjures some notion of an individual or their work: the inimitable work of Leonardo da Vinci, concretely; or, abstractly, a medical doctor, nuclear scientist, accountant, classical violinist, or champion chess player. When one studies how someone becomes an expert, numerous theories of how one gains knowledge, skills, or practices can be found across various scholarly fields. In other theories of expertise, the social nature of the individual expert is examined, asking how attribution of status helps construct the expert. Normally these experts are situated within disciplines, specialties, or professions where their knowledge or skills strikingly surpass the skills of others working in their area. Recognition of this individual's extraordinariness by their peers or publics is another factor in how one's status as expert is sanctioned. The nature of experts or expertise has changed over time, but the belief that some people have greater knowledge or skill than others in some areas is well established. For decades researchers have taken this understanding and tried to establish psychological and social rationales for differences in knowledge and skill in expert performance.

A rhetorical account of expertise may appear most obvious in a discussion of the artistic or socially constructed and sanctioned ways that experts attain and maintain their status—a study of their ethos, or credibility and ethical comportment toward one's audience, for instance.[11] But it is the capacity for deliberation and judgment that are most crucial to the enterprise of expert knowledge and performance. Deliberation and judgment further complicate the already complex notion of expertise, but rhetorical studies offer a substantial body of thinking on the issue. Moreover, rhetorical studies offer a nimble theoretical framework. Hartelius, in her 2011 work *Rhetoric of Expertise* (164), explains, "Approaching expertise as a rhetorical construct releases us from some of the constraining dichotomies that seem to plague the topic.... We can use a rhetorical hermeneutic and begin a productive investigation with the assumption that style *and* 'real' knowledge are not only integral but inseparable." Further, examining both the individual and collective nature of ethos in expertise provides a rich understanding of its dynamic nature. Finally, and crucially, understanding that expertise attends to matters of audience is central to a fuller understanding of expertise.[12] Research in rhetorical studies also offers bridges to other fields studying expertise, including the psychological sciences, and through comparative work helps further explain how we cultivate expert capacities and enact expertise through expert performance. Contemporary rhetorical studies, following several gossamer threads back to antiquity, further offer a cognitive[13] account that is impressive in its investigative power. Indeed, as we will see, lessons from antiquity and the medieval period concerning the art of memory can advance rhetorical understanding of the operations that underlie our inventive, recollective, and experiential mind, allowing the cultivation of expertise. Indeed, these accounts of the rhetorical habits of mind advance a pragmatic, situational model of expert performance. Key to these habits is phronesis, a concept intricately connected to the idea of rhetoric. When we understand expertise as requiring phronesis, we must understand expertise as an ongoing engagement by some individual situated both socially and also by the experiential aspects of their being.

Understanding the role of phronesis in conceptions of rhetoric can help elucidate why conceptions of phronesis are also illuminating in the discussion of expertise. Self (1979) provides an important argument for how and why phronesis is a key concept in articulating Aristotelian conceptions of rhetoric. Explaining that Aristotle's ethical understanding of rhetoric is quite well developed, Self reasons that to understand the ethical nature of his rhetoric, one must look to Aristotle's ethical arguments in both *Rhetoric* and *Nicomachean Ethics*.[14] Here Self (1979, 135) provides a succinct account of this argument, writing, "Rhetoric is an art, *phronesis* an intellectual virtue; both are special 'reasoned capacities' which properly function in the world of probabilities; both are normative processes in that they involve rational principles of choice-making; both have general applicability but always require careful analysis of particulars in determining the best response to each specific situation; both ideally take into account the wholeness of human nature (rhetoric in its three appeals, *phronesis* in its balance of desire and reason); and finally, both have social utility and responsibility in that both treat matters of the public good."

A key capacity here for both the art of rhetoric and the enactment of phronesis is the ability to deliberate on a particular situation. Self reminds us of the important connection between conceptions of deliberation in Aristotle's *Rhetoric* and *Nicomachean Ethics* with the use of the term *bouleusis*. *Bouleusis* is used in both texts, describing, in *Rhetoric*, "the process of deliberation," and in *Nicomachean Ethics*, "the faculty of the man of practical wisdom" (Self 1979, 137). Phronesis, or prudence,[15] is reformulated for techno-scientific modernity and scientific and technical expertise by Danisch (2010). Prudence, Danisch reminds us, is a concept that has evolved along with cultures and a modern conception of prudence can be located in scientific thinking, which is a "special case of practical reasoning" (189). Majdik and Keith (2011a, 2011b) also provide a framework wherein a conception of expertise is relational to problems or tensions, rather than individual capacities. Articulating expertise in this way affords what they describe as a "practice-centric view of expertise" (2011b, 276; see also, especially, 279). In this view, enacting expertise is measured by one's ability to articulate a socially grounded and recognizable rationale for one's decision-making. Further, Hartelius (2011, 171) discusses the importance of prudence in understanding the relationship between the particular and the universal, and the ability of experts to situate their expert or specialized knowledge in relation to "'big picture' significance." It is notable, here, that such a conception of "traditional expertise" takes on a distinct form, notably "serious," from what Hartelius (2020) articulates as a "gifting logos" or rhetoric of expertise in the digital commons that is keyed into playfulness, *copia*, continuousness, and the common(s). For the purposes here, the discussion of expertise is chiefly preoccupied with the "serious" conceptions of expertise, but Hartelius's attention to expertise in the digital commons makes a critical move to emphasize the productive epistemic contribution of expertise as making and as gifting, which goes a long way to think about futures for the "serious" or "traditional" forms of expertise, too.

For Danisch, an understanding of expertise as a kind of prudential thinking, or phronesis, is situated within the "risk society" that Beck (1992)[16] has argued has been constituted by techno-scientific modernity.[17] For the purpose of discussion here, this is an especially helpful formulation. Despite efforts to alleviate uncertainty[18] through science and technology, new forms of the uncertainty arose. For Beck (2009), and for Danisch, the conditions of risk society demand global democratic engagement rather than a retreat into an impossible fantasy of cloistered expert knowledge and the scientific ethos as one concerned only with facts.[19] But the situation is somewhat more complicated than rebuilding the *agora*. Rather, risk has reconfigured the values of a democratic public, adding the value of security to democratic values of equality and freedom (Danisch 2010, 182). Current configurations of expert-public relations frequently attempt to situate experts as being in possession of objective knowledge, but it is this positioning that itself fails, Danisch explains. Indeed, because of the positioning of objective knowledge, it is knowledge that has little application to any problem that requires a solution. Further, the possession of such proclaimed objective knowledge is predicated on an antagonistic relationship with publics. The relationship frames the public, the nonexpert, as less rational or even irrational, a problem only to be solved by the import of expert knowledge to remedy such unscientific thinking. Antagonistic relationships lack the goodwill toward an audience that would reasonably dispose them to hear an expert's arguments and forecloses conversation between the expert and the audience.[20]

Refiguring what falls under the domain of rhetoric is predicated on what we understand to be certain or probable knowledge. However, for Danisch, the move we must make is to understand that in a risk society, science and technology produce not certainty but probable knowledge. Thus (pace Aristotle), scientific and technical subjects are firmly in the domain of rhetoric. For Danisch, the rhetorical power of twentieth-century experts is imbued with authority. In the twenty-first century, the rhetorical world of expertise has lost some of its authority (Nichols perhaps would not be surprised by Danisch's claim). Rather, Danisch (2010, 188) argues that the "risk society thesis shows that the conditions of possibility for returning judgment and authority to common citizens are now in place, but this has not happened yet."

Prudence might help us remedy this situation. Prudence, Danisch (2010) notes, is historically contingent and, that being so, one might ask, as he does, "What would a scientific and technical form of prudence look like?" or "How would we train citizens in the cultivation of a scientific prudence?" Beginning to formulate an answer involves a conception of prudence aimed at "understanding the ways in which scientific work is itself deliberative and conditioned by uncertainty and controversy" (188). It is not enough, however, to theoretically understand this proposition, nor for scientists alone to understand. Rather, the public also needs to understand this form of practical wisdom that we ascribe to scientific and technical reasoning (189). Most crucially, such reasoning would illustrate how science and technology advance "moral agendas" and ensure that "prudential citizens would have the capacity to read the morality of techno-scientific rationality" (189). A kind of scientific expertise articulated through the lens of phronesis is an important antidote to the seeming confusion concerning the role of expertise today.[21]

Following Danisch, I am interested in the role of phronesis in scientific and technical reasoning as a special application of phronesis and am principally preoccupied with the expert's capacity and need for such practical wisdom. Rather, because expertise foundationally operates relationally, I am interested in understanding how the capacity and need for phronesis shapes expert thinking and doing. I am also interested in how this concept can call us to evaluate our experience, its limitations, and our commitments to others in an ongoing way. That is, attention to how phronesis itself is an ongoing process that, in being socially situated, demands recognition of its limitations. For example, cultivating this capacity should attend to where its socially situated sensibility of "the good" might, in fact, replicate, for example, sexism, racism, antisemitism, or ableism. Katz (1992, 1993) notes the always socially defined nature of phronesis itself, as reasoning, and of what is good, as a *telos*[22] of phronetic reasoning and action, warning specifically of the dangers a phronesis grounded in an ethic of expediency holds. Scientific prudence is not categorically different from other forms of phronesis in this way. Danisch's argument for the need of prudential citizens to understand technical and scientific arguments, however, follows in an era where our social expertise is so shaped by these discourses. Katz (1993, 45) illustrates the dangers of modern technical and scientific discourses and reasoning in his "The Ethic of Expediency," where he explains that phronesis is, like other forms of knowledge, socially situated and created. It is important, then, when considering phronesis as a capacity for deliberation, to note that deliberations occur in the already shaped community and culture. That is, phronesis is socially defined.

Even where we might identify and acknowledge the limits of our ethical deliberation, situations may make the best course of action impossible. At the time of completing this book, we globally face a continued pandemic that has demanded daily ethical decision-making that illustrates further the challenges to each person's phronetic capacities. Wasserman (2020, para. 4) notes how the ethical conversations about the COVID-19 pandemic, which include alarming crisis-based decision-making such as choosing who should be provided life-extending care where such care is limited, "narrow our ethical imagination." Ethical decisions made prior to the pandemic have dramatically shaped the outcomes, as is evidenced by, for example, the disproportionate impact of the pandemic on racialized communities, in Indigenous communities, the significant losses of life in long-term care homes, and the economic impacts on women, especially women of color. Katz's (1992, 1993) caution of the ethic of expediency echoes in these ethical failures where economic and political expediencies drove decision-making for decades or longer. Phronesis, in the face of these social inequalities, can describe the capacity to challenge the ethic of expediency, but must also be guarded against the zero-sum ethical deliberations Wasserman cautions against in the current pandemic and, broadly, how we live together. Indeed, as phronesis is a socially constrained mode of ethical thinking, awareness of current limitations and biases is critical.

Insights from Virtue Ethics

The ancient traditions of virtue ethics can illuminate the "virtue" of phronesis and how someone might cultivate the capacity for such a virtue.[23] Virtue ethics today can be situated among three major schools of thought in normative ethics, each of which has numerous subfields and permutations.[24] Virtue ethics are concerned with one's character, how one becomes a more virtuous person, and there are pluralities of virtue ethics in Confucianism, Buddhism, Christianity, Judaism, and other traditions. In the ancient Greek tradition characterized by Aristotle, becoming virtuous was crucial to obtaining *eudaimonia*, a kind of fulfillment or flourishing of a person. Indeed, individuals, in this model, must work to habituate themselves, to become virtuous. While a person may have some natural virtues, these are uncultivated and might lead us to do the wrong thing. The ongoing cultivation of virtue is central because it will allow virtuous individuals to know that they have done the right thing, to not worry that they have misstepped. Thus, the virtuous person might experience *eudaimonia* (i.e., be fulfilled). The ideal here—occasional lapses in judgment are not understood as a condemnation of someone's character—is that someone will be in harmony with doing good. However, the realized, situational, and changing social world we inhabit likely means there will be some tensions as we work toward this ideal, but perhaps we could say we finds some harmony even in those moments where we have made mistakes, allowing us to thus continue habituating ourselves to the good.

The changing conception of the good is one attribute that underscores the rhetorical dimensions of a virtue-based approach. Indeed, Blumenberg (2020, 180), writing on the rise of Academic skepticism and its consequences through to nihilism, argues that an "ethics that takes the self-evidence of the good as its point of departure leaves no room for rhetoric as the theory and practice of influencing behavior on the assumption that we do not have access to self-evidence of the good." Thus, we might understand such a rhetorical virtue-based approach to require situating all phronetic reasoning as part of the reasoning itself, a kind of reflexive phronesis. But this moves past where we might begin, with an Aristotelian conception, from which we can traverse virtue-based ethics vis-à-vis Greco-Roman rhetoric, and beyond. The departure point is owing to my own training, work, and limitations, and I do not have the expertise to engage in significant comparative work (see, on the challenges and significance of the undertaking, Sim 2007).

For Aristotle, there are both intellectual and moral virtues. Moral virtues tell us about habits by which we might become virtuous by finding the mean, or the middle ground, among vices of excess and deficiency. It is, however, an intellectual *and* moral virtue—phronesis—that is especially important to developing our capacities. Phronesis allows us to deliberate on a problem and make the appropriate judgments, including in matters of moral decision-making, in habituating ourselves to become virtuous.[25] It is not enough, however, to only live ethically without regard to what is true, the purview of intellectual virtues, and thus the two are interconnected. Already we can identify in this relationship how phronesis and its capacities are not distinct from that which we might place in the scientific and technical domain. The importance of this virtue even in these domains is clearly manifest through the importance of rhetorical activities, particularly in this risk society where we ruminate on the seeming death of expertise.

To cultivate virtue is to practice habits, not mere routines, that allow the parameters of a situation—its very existence, even—to be identified, to understand the circumstances within which one might act, and to determine through reflection and deliberation the appropriate action that demonstrates our prudence. We accomplish this rhetorically when prudence moves into the public sphere. Duffy, Gallagher, and Holmes (2018) demonstrate the value of virtue ethics to a rhetorical approach and to our contemporary discourses. They explain that the virtues associated with a person of good character (including virtues of "honesty, accountability, generosity, intellectual courage, justice, and others") living the good life can offer "an alternative to the toxic discourses of post-truth, alternative facts, and other practices of disinformation" (323). Contrasting with a prescriptive moral framework, Duffy et al. argue that a virtue ethics approach helps us respond appropriately, too, to oppressive situations. They explain that "skepticism, righteous anger, and resistance are also virtues," which are crucial to responding to oppressive situations (323).[26] Further, because a virtue-based morality "is not guided by codes or rules," the virtuous person "knows which virtues to enact, in which settings, for which reasons" (323). These are ideals of one's character, ideals we work toward through processes of habituating, and the individual is not required to reflect ideals but to strive toward them. For those familiar with the research on expertise, this should be a familiar refrain. Experts, too, require a process of habituating and are always striving toward expertise rather than being in possession of it per se.

Expertise and expert status align with this model because they are rhetorical activities—that is, activities that operate not only within one's own reflection on truth or with one's philosophical or scientific or technical development alone. Expertise and, perhaps more overtly, expert status are enactments of technical knowledge within a community that has some stake in the truth claims. Although it may seem that a virtue perspective is preoccupied with the individual, as Bergès (2015, 114) notes, in the Aristotelian tradition and in Platonic conceptions the character of individuals is "regarded as operating within a community. Virtue is seen as that which enables us to perform our function well; hence, a part of flourishing depends on being part of a city." Maimonides, too, notes the foundational essence of virtues as our relation to others, writing of the third kind of perfection a person can achieve—moral perfection—that "all moral principles concern the relation of man to his neighbour; the perfection of man's moral principles is, as it were, given to man for the benefit of mankind" (*Guide of the Perplexed* 3.54).[27] Among the virtues, and perhaps mostly directly imported to rhetoric by Aristotle, phronesis as a governing virtue affords an important way into the discussion of experts and the role of virtue. Gage writes of a *rhetorical* conception of phronesis, that it may be understood as "*the ability necessary to make informed judgments about the whole rhetorical situation one is in relative to one's own beliefs and needs and the beliefs and needs of others, and about the selection and disposition of rhetorical means to adequately address the exigencies of those situations*" (2018, 329, emphasis in the original). Underlying the importance of phronesis is always, Gage explains, "the absence of certainty" (329). Absence of certainty is a notable feature of situations where we require expertise and experts. If we understand expertise merely as mastery of some scientific knowledge (episteme) or skills (techne), we remove the sense of the social, real-world application.

Further, before even integrating another form of knowledge, we must see the importance of integrating such theoretical knowledge and also skills, what we might call the *knowing-that* (e.g., the theoretical knowledge, scientific knowledge, or episteme) and *knowing-how* (the skills or techne).[28] Skills are composed of a complex arrangement of techne, or craft, along with theoretical knowledge, as Annas (2011) explains. One cannot merely know-how (have a skill) without knowing-that (have some theoretical understanding) and be said to properly have a skill (techne); rather, know-how alone might rather be described as a knack (*empeiria*). This is not to suggest that an anti-intellectual commitment to *know-how* (as "instinctive skill") is a path toward truth (White 1962, 468; see also, especially, 461). Rather, this is to distinguish between modern vernacular conceptions of skills when dismissed relative to "knowledge" as meaning something closer to *empeiria* than techne. To understand a skill as something less-than, such as in these uses, would be to reduce and distort Aristotle's conception of phronesis as *like* a skill (as techne). Further, skills are inclusive of forms of knowledge where we may know more than we can say, following Polanyi's (1966) conception of tacit knowledge, and are also inclusive of kinesthetic modes of knowledge, which offer a provocative way into exploring—reducing or dissolving—subject-object relations (see, for instance, Harper 1992, 133). Knowing-how without knowing-that amounts to an "inarticulate practical knack, an ability to manipulate the world which is not at a sufficiently rational level to be judged epistemically," as Annas writes (2006, 290). Allowing for a perspective where knowing-how is divorced from knowing-that, we are able to accept that there is "no such thing as practical expertise, only knacks—that there is no significant difference between the inarticulate practitioner and the expert in the field" (290). "This," Annas argues of a position where knowing-how does not involve in anyway knowing-that, "is ridiculous" (290).

Understanding the relationships of theoretical knowledge and skills in moral knowledge is illuminating in expert knowledge because it helps us chart the similar relationship between knowledge and skills in expertise. Further, this understanding also clarifies why moral aspects of practical knowledge must be considered. When theorizing moral knowledge as or *like* skills, there are certainly features that distinguish such skill from those commonly associated with expert practice. The distinguishing feature is that moral knowledge-as-skill is a global skill in living one's life whereas what we could conventionally think of as expert skills are quite local in their application. However, in cases of both moral and technical knowledge, practical wisdom is a capacity that can help develop a systematic understanding of a subject where there were once "piecemeal" beliefs accounting for the operational framework of understanding. Socrates's method is principled on such an effort to move from piecemeal, belief-based understandings to more systematic understandings. A range of expertises also require this kind of systematic understanding, from building a house to, Annas explains, the understanding of moral forms in the Platonic tradition. Within such a formulation of knowledge, we might see how it cannot be divorced from expertise but also how it does not itself constitute expertise. But this is not, in fact, inconsistent with the model of expertise developed here with respect to moral knowledge, but rather seems to be restricted to models of knowledge as episteme and perhaps some form of practical knowledge with respect to experiences. Further, "skills" in the contemporary sense are not well distinguished from knacks (*empeiria*), but for Aristotle and others in Greek antiquity skills (techne) required intellectual articulation (Annas 2011, 19–20). Moral knowledge may seem different, as Annas (2006, 290) speculates, because "coming to understand a moral Form is harder than the other cases." Consider, as Annas does, how moral questions are likely to pose more complex challenges than the challenge of, say, learning something about electronics. Critically, however, Annas argues that this discrepancy in difficulty is "in itself is not reason to deny that both are examples of practical knowledge" (290). Ultimately, Annas makes clear that moral knowledge and expert knowledge are forms of practical knowledge. But it is not enough to say this is so and these two forms of practical knowledge are alike in certain ways that provide us some understanding of the other. Such is the case because where we equate experts with expertise, and expertise with skill—inclusive of knowing-that (episteme) and knowing-how (techne)—we are left with a rhetorically incomplete understanding of experts and expertise. Practical knowledge offered through phronesis, combined with practical knowledge as techne, provides a vital intersection for understanding about experts and expertise in a rhetorically complete way.

Such a sensibility about moral knowledge as practical knowledge is a recurring theme in developing a rhetorical account of phronesis as a moral comportment that experts must acquire. Indeed, the requirements for phronesis include technical capacities and, crucially, the ability to deliberate. Aristotle's account of deliberation as a technical and moral activity in phronesis is treated in the next chapter, but here the more pressing matter seems to be a contemporary orientation to understand desire as a foundational motivational force in philosophical studies. Annas (2006, 292) refutes the question of desire in the case of experts: "Experts deliberate about the objects of their presence by pain and frustration until they are fulfilled. . . . Experts deliberate about the objects of their expertise, not about how to fulfill their desires (of course they might do the latter, but not in a way relevant to the exercise of their expertise)." Rather, it is a state of enjoyment of the action that seems relevant, and Annas links experience of performing virtuous actions to Csikszentmihalyi's flow, a kind of complete immersion in experience and performance (see Csikszentmihalyi 1990, 1996, 1997). Experiencing virtuous action as the virtuous person is likened to this state. What I suggest here, contrary to Annas's argument, is that experience of good reason for the expert, like the virtuous person's performance of different virtues, has a similar and possibly overlapping phenomenology.

Diverse ways of knowing provide insight into complex forms of knowledge and expert understanding, including those tacit dimensions affecting the seemingly effortless performance of experts. A critical concept for understanding tacit dimensions is the idea of phronesis, which allows for a complex operation where forms of epistemic knowledge, experience, and skill are enacted through an ethically framed capacity. Those ethical dimensions that emerge in this exchange of expert and audience are enacted through ethos, in concert with an audience. In the former understanding, practice begins to provide some insight into how a novice acquires the required knowledge and abilities or skills along their path to being expert. In both accounts, many researchers have noted how knowledge and practice function together to provide experts with models that allow them to more effectively, for instance, "parse" knowledge (Bereiter and Scardamalia 1993, 37) or "chunk" information (Chase and Simon 1973). Being able to parse information allows experts to better assess and respond to some exigence.

Even in those constrained, measurable situations such as those studied through the cognitive approaches of Chase and Simon, such as Elo-rated chess matches, the variables introduced in the complexity of a game, including the opponent, their strategies, and so on, all require experts outperform even some of the strongest competitors. "Winning chess matches (or any kind of game)," as Douglas (2018, 98) writes, "is not going to help us grapple with the kinds of complex problems society faces." We require, that is, that experts can operate in uncertainty, otherwise the measure of expertise is extremely low. More challenging moments, the moments that push the limits of our knowledge, are better indicators of becoming expert. To operate in those uncertain, challenging moments, we require phronesis. Gage (2018, 329) explains that we can characterize phronesis as the "will to act (rhetorically) with confidence but without certainty, which entails taking the risk of being ineffective or unbelievable despite one's best efforts," and do so within an effort to do good. Expertise, then, is not simply a matter of acquiring some knowledge and practicing some skill, but, crucially, of *applying* knowledge and skill to some problem, some situation, and doing so with good intentions. Even when considering the role of theory, the knowing-that, its application is not only to further knowledge building in singularly disciplinary contexts. Fan (2020, 40) powerfully argues, for example, that "the point of theory in times of crisis is to be found within crises themselves—in action in street protests, hospital wings, and legal institutions around the world." It seems, then, sensible that we might understand the role of phronesis as that intellectual capacity to deliberate. The question of good intentions in doing so, however, requires further consideration.

In its moral constitution, phronesis requires that expert performance would be done with good ends, not merely achieving good ends. Such a distinction is important because it puts the character of the expert at the center of the act. Here we find a seemingly difficult problem, because certainly someone can have highly specialized knowledge or skill, as well as some ability to judge situations, but ultimately serve ends we might, as a society, a public, say are distinctly *not good*. Consider, for instance, a highly proficient doctor who decides those patients who should be mended and those who might, even in the presence of viable treatments, be left to die. In medicine, the built-in ethical norms suggest that physicians acting in this manner would not be expert insofar as they violate the commitments of their oath, failing to understand their social role as designated medical experts. In normal operating situations this may be the case, but the COVID-19 pandemic demonstrates how this question is vastly more complex than one might hope and entails situations and decision-making preceding the pandemic. As hospitals reached capacity in the spring of 2020 and physicians made impossible ethical decisions about whom they would try to save and whom they could not help, the ongoing requirement and limitations of phronesis are painfully demonstrated. Physicians, in their duties, too, would succumb to the disease. Here the situational, moral questions these experts are tasked with making on an ongoing basis is even more starkly illustrated than the already difficult life-and-death decisions physicians sometimes make, and without definitive answers.

In another domain, we might look at software engineers who build new technical products that reproduce racism (see Noble 2018). Their technical ability, either theoretical or practical applications of coding, allow them to act in a seemingly expert manner, but what they produce may, in fact, offer little in the way of what is good. Instead, the supposed expertise, in fact, perpetuates forms of violence and oppression that are inexcusable. Although the tools may seemingly have some functionality that eases the lives of some, it comes at a cost to others. Even in those cases where racism or sexism might be less embedded (it is unlikely to be absent) in the technologies, many fundamentally operate in such a way that convenience is afforded by the cost of privacy—we might say, often, an unwitting exchange for consumers (see Fernback and Papacharissi 2007). But is the highly skilled, well-compensated tech wizard in fact an expert if lacking a socially responsible and responsive comportment? Socially, we might very well say yes; however, Aristotle provides us a good reason to say no. These individuals, rather, have a knack (*empeiria*) that allows them to excel at their occupation, but they do so in part through ignorance (or, in some cases, he might caution, deception). They do not act in service of the community,[29] but rather the bottom line. Here the community makes a social decision of who is expert, whom to trust, and this is certainly debatable. However, the status of expertise, its purported death, and rejections of experts suggest there is, even in these highly technical domains, some sense of a social contract among experts and publics that has failed. That is to say that expert status *is* a social agreement. The social agreement, I wish to add, also means that sexism, racism, and other forms of prejudice have an impact on who is viewed as an expert, and this can vary greatly across contexts, which I discussed at length in earlier versions of this book, but which, indeed, require their own substantial treatment that I could not adequately explore in the provisional account offered here. Expertise, too, is a social agreement insofar as it is not mere technical competency, but a form of knowledge that others require to make informed decisions about their lives, which makes violations of such an understanding dangerous. Expert status and expertise, thus, are subjects to which rhetoricians and professional communicators can contribute important insights. As Walwema (2020, 36) explains, scholars studying rhetoric and professional communication can illuminate the role "expertise can play in fostering the trust that enables the public to make good decisions." Rhetorically, the deliberative genre (forward-looking) of expert performance, often inexorably linked to the forensic genres (backward-looking) with which expertise is partially constituted, demands of experts phronetic capacities, inclusive of moral deliberation. A rhetorical understanding of expertise is not simply *knowing-that* (episteme) and *knowing-how* (techne) but *knowing-why* (as enacted through phronesis, practical wisdom, and based on its underlying capacities).

Speaking with the Experts

This book uses rhetorical approaches to understand the practice and relational-based theories of expertise in the multidisciplinary literature. In doing so, the central argument of the book is that a rhetorical account of expertise shows that both individualist and relational models of expertise explicate various aspects of the phenomenon of expertise, and that both are illuminating for a comprehensive rhetorical understanding of the concept. Further to this point, a rhetorical approach relying on both individual practice and attributional-relational features is required. Rhetoric is an inventive art, one that shows us how seemingly rote practice and memorization is, in fact, more inventive and, crucially, fundamental to enacting expertise in each situation, along with identifying the recurring situations and associated expectations for performance. As an inventive art these situations are pregnant with affordances and *kairotic* moments, and constrained by social norms and conventions, expert knowledge and patterns of thinking, and audience expectations; further, they are in each instance, new while invocative of past recurrences.[30]

Within this framework, the audience is recalled, often absent in accounts of expertise, and so too are what we might call cognitive aspects, including the affective, ethical, social, and their intersections. The continual improvement of expert performance, as documented thoroughly in the psychological sciences, illustrates this case. Consider the high performing athletes of the 1940s and today. Today's athletes far outpace their earlier counterparts due to the evolution of their training as well as the situational features of expert athletic performance with respect to audience. The evolution of experts and expert training reveals audience as competition, audience as judges, audience as consumers of sport, et cetera. Rhetoric offers a complex theoretical framework that allows for contingencies, characters and credibility, socialization and socio-cognitive apprenticing, tensions between stabilization and change, and cognitive wetware in a formulation of expertise. Unfolding some of the complexities that constitute expertise and expert status is one of the chief goals of the present work. Aligned with this objective, the research program that provides expert insight from surveys and interviews[31] for this book sought to understand how experts evaluate others' expert status in multidisciplinary settings.[32]

Multidisciplinary research teams offer an interesting site to explore expertise because they have many kinds of experts involved, and the complexity of assessing expertise is considerable. The rationale behind examining multidisciplinary teams[33] is that participants cannot measure only competencies or proficiencies in a specialized area as a marker of expertise. Consider, for instance, a multidisciplinary academic research team comprising an ecologist, a computational modeler, an anthropologist, and a historian. Each of these researchers have some expertise—in terms of capacities and attributed credibility—in their areas of specialization. Although there may be some overlap in interest or specialization, or even perhaps methods and epistemological commitments, each researcher has a different configuration of expertise within their disciplinary home. Given that each expert will have limited abilities to fully assess the expertise of any other expert based on demonstrations of knowledge and skills within a specialty, the group of experts is likely to develop other measures of expert status and credibility. That is, if it is impossible to assess highly specialized skills required of an expert in each specialty, it is necessary to otherwise calculate the likelihood of one's colleague's capabilities. Because the rhetorical dimensions of expertise are negotiated and adjudicated by various kinds of professionals, it seemed prudent to sample a variety of these professionals to learn about what they assess. Speaking with professionals about how they understand their own expertise, and how they believe they became experts, provides fascinating insights, too. Surveying and interviewing self-defined experts in multidisciplinary teams provided the basis for this research.[34] The research program focuses especially on members of multidisciplinary science, technology, engineering, and mathematics (STEM) teams. My interest in multidisciplinary teams quickly necessitated an expanded range of participants, and my research program might ultimately be said to include members of science, technology, engineering, arts, and mathematics (STEAM) teams and citizen scientists. Not wanting to conflate expertise with professional status, citizen scientists are an important group to include because they are everyday people who participate in scientific research.[35]

Using a survey,[36] participants identifying as professionals and experts[37] were asked a range of questions, including about their job title, field of study, degrees, years of postsecondary study, and years of professional experience.[38] Participants were also asked about their confidence in assessing experts in their own area and others, words they associate with an expert or someone who has expertise, and how they identify collaborators. Semi-structured interviews[39] were also used and participants identifying as professionals and citizen scientists[40] were asked to tell us how they defined expertise, how they became experts, and how they determined if the individuals they work with in these teams are experts.

Overview of Chapters

Beginning in the first chapter, the tradition of virtue ethics and its implications for a theory of expertise are explored alongside current social theories of expertise. In this chapter, the idea of how expertise both requires moral knowledge and is a continual process of being is advanced. Virtue ethics is essential because it provides not only a model of moral philosophy, but one tied to the development of Aristotle's theory of rhetoric. Further, the tradition helps us trace some of the individual, agent-based aspects of expert capacities that cannot be located only in the activity of "expertising," for example, but in the cultivation and habituation of mental operations. In chapter 2, research in rhetorical studies of science, a subfield of rhetorical theory, illustrates how expertise is indeed a rhetorical act. Drawing on concepts of ethos, character, as well as practices of expert communities, this chapter situates rhetorical studies in broader studies of expertise and illustrates that rhetorical studies recenters two important aspects of expertise, the audience for expertise and the moral character of the expert. Chapter 3 examines psychological theories of expertise, notably studies of practice and memory in expertise. Rhetorical theory powerfully articulates the capacities that allow one to cultivate what is here referred to as expertise through the study of memory, which aligns with but also complicates contemporary psychological studies. Memory and its relationship to prudence during the medieval period helps illustrate how expertise is more than mere cognitive capacity, but an enactment and comportment of knowledges within an extended ethical framework. An ethical accounting of expertise allows us to examine the notion of someone who attempts to do right and understand the best solution in a situation without preoccupying ourselves with specific, singular instances of ethical concern. We see this in medieval memory through to de Groot's foundational studies on chess masters and contemporary cognitive studies. All of this, too, is grounded not in the kind of positivist, super logical, and rational individual, but in matters of emotion, individual experience, relations to others, et cetera. What that tells us is important, too, for how we think about and communicate with audiences, because the way rhetoric permeates even expert thinking provides a bridge to the nonexpert by way of virtue, goodwill, and good reason—ethos, but ethos as deeply ethical, community based, and relational. Experts, then, are characterized not only by those cognitive capacities, but as situated rhetorical actors.

Throughout the first chapters, experts explain in a series of vignettes how they believe they came to be experts. It is not surprising, given the complexity of the task, that different experts may have different perspectives on how they became expert. Subjective experiences are reported; indeed, they must be subjective when we ask someone to reflect on potentially decades of learning, experience, practice, and argument, as well as material or social barriers, personal challenges, and so on. Because learning is an individual experience, the responses given by participants provide examples to apply the complex theories that underlie the rhetoric, philosophy, sociology, and psychology of expertise. The stories offered by our participants are smart, funny, and help unravel in multiple forms how we cultivate expertise.

In chapters 4 and 5, a thematic analysis brings together a series of lessons offered by participants in their rich and thoughtful responses. In chapter 4, exploring survey and interview data, the voice of self-identified experts articulates how they conceive of expertise and how they assess other experts. Using data from survey responses as well as excerpts from interviews with self-identified experts, a complex account of expertise emerges. Chapter 5 continues the investigation with insights from citizen scientists. Citizen science normally describes two broad ranges of activities. First, there are those citizens who are enlisted to participate in scientific research as designed and governed by professional scientists. In another model of citizen science, everyday people initiate the research and govern its progress, rather than being directed by scientists. Important in their responses is an expanded understanding of expertise and experts beyond professional confines, extending to those uncredentialed but working within a specialized area. The concluding chapter of the book distills the case for a rhetorical approach to expertise, emphasizing the importance of moral knowledge as phronesis in the cultivation of one's character and expert abilities, in relation to one's goodwill toward expert, less expert, and nonexpert audiences. Distinctions between how one's expert status operates, its ethotic qualities, and the conception of expertise remain a preoccupation. Ultimately, the book argues that expertise is the enactment of knowledge and skills, through practical judgment and practical wisdom founded on integrated experience and, critically, through an ethical framework relational to one's audience, and most applicable to the situation or problem one faces.

1

Habituating Expertise as Rhetorical Act

Expertise is cultivated as one moves from novice to expert thinking through habituating oneself to act expertly. We know one might fall from the status of expert, no longer holding the exceptional knowledge, skills, and deliberative capacities we have come to understand as expertise, for the duration of an entire life. Athletes are illustrative, given that most will, eventually, meet a barrier to their ongoing expert status on the basis of physical decline associated with aging. However, such athletes may, in another capacity, continue to become expert, for instance through becoming trainers or coaches. This may be true of other domains where expertise may be lost over time, owing to not only cognitive or physical reasons but also emotional exhaustion. We might think here of health care professionals who experience serious burnout. Whatever the case for the loss of expert capacity, this may indeed be a transformation rather than a mere ending, but in either case the impermanence of expertise is notable.

In this book, expertise is not limited to the professions, and is never a complete process. The participants who were surveyed or interviewed for this project provide key insights to the varieties of expertise and its processes. Because their perspectives are varied, their insights can help illustrate the theory developed in this and the following two chapters. By way of vignettes, these participants will illustrate the complexities of expertise. Consider, for example, Grey. Grey is a researcher holding a doctoral degree in the life sciences. Although we would easily categorize Grey as a professional researcher, Grey is also a citizen scientist who builds computational platforms. Although Grey has considerable experience and holds degrees in science, they do not identify as an expert. "To be an expert," Grey tells us, "is something that you can never reach. But every time that you go outside, you are getting closer and closer." Experts, that is, have "experience." When asked about how they became an expert—or are, at least, on a path toward expertise—Grey commented, "I think your question is related to being a citizen scientist more than how to be a professional researcher because, of course, you want to be a professional, you have to go through the career, postgrad, and whatever." Interestingly, this comment highlights the relatively clear path one can take in established disciplines on their path toward expertise, largely guided through institutionalized mentorship and guidance. For citizen scientists, however, the path can be more varied. Grey suggests searching online to find an already established project is a useful strategy. Even once one finds a project, there is still considerable work to do to gain some competency, let alone expertise, that aligns with the work of a project. To begin the path toward becoming expert in some area, Grey explains that one should first acquire some basic knowledge through reading and then get out in the field.

Doing the work of citizen science, however, also involves working with others, and Grey explains that many people working on citizen science projects want to help new people: "Let's say they are not so confident with identification of the species. Normally, there is no problem. There are people behind, looking for your observation, so don't be afraid to not understand your ecosystem. You can go outside and look for it." There is an interplay between knowing-how and knowing-that which is important to theorizing the forms of knowledge produced in citizen science. Identification guides and the like provide the basis of knowing-that, the more formal sorts of knowledge that are well supported and codified into a body of scientific knowledge. The knowing-how of species identification, however, is challenging because the guides only provide general heuristics for identification, but there is much nuance and difficultly in its actual practice. But the know-how of such identification vis-à-vis scientific classification systems is impossible without knowing-that; this is to say, know-how as techne, not mere empeiria. But there is more to gaining expertise, Grey says, than only techne, or even episteme. In citizen science projects, perhaps more clearly than professionalized scientific efforts, there is space for a discourse of motivation, of passion, of engagement with the subject. Such an idea alone will not result in expertise— for the reasons outlined by either the practice-based psychologists such as Ericsson or the other psychologists of expertise, we will later learn, who see importance in cognitive features—but without it, expertise is, perhaps, impossible. Grey explain that when beginning to participate in a citizen science project, "having a good time and feel[ing] fine with the effort you can give" can help "you start to increase your expertise." "It's difficult for me to imagine splitting your expertise level and your willingness to participate," Grey remarks. Grey's advice, then, when imagining how one might become expert is to "give the proper importance" to motivational aspects, saying "if you love what you do, and if you are encouraged to learn, and to be nice, collaborative, you will be absolutely an expert."

In this chapter, the very idea of expertise is explored, first through the lens of virtue ethics and then through social theories of expertise. The important processes of habituation are a common theme among the kinds of practical moral knowledge described in the virtue ethics tradition and in the fields of research studying expertise and its attendant theoretical and practical knowledge. A virtue ethics tradition, drawing initially from Aristotle as connected to his rhetoric, also provides a rough theory of knowledge to begin untangling the complex forms of knowledge required for someone to become expert, as well as the relational-rhetorical aspects that allow one to obtain expert status.

Aristotelian to Contemporary Virtue Ethics and Expertise

Phronesis is among the intellectual virtues described by Aristotle in *Nicomachean Ethics*. Aristotle set out a tripartite model of the human soul,[1] which is constituted by the vegetative, appetitive, and rational faculties.[2] Intellectual virtues are those governed by the rational part of the soul (see Aristotle, *EN* 1.13 and 6.1). Book 6 addresses five varieties of intellectual virtues (or excellences), including *episteme* (scientific knowledge), *nous* (intelligence or, following Zagzebski 1996, "intuitive reason"), *sophia* (theoretical wisdom), *techne* (art or craft), and *phronesis* (practical wisdom) (Aristotle, *EN* 6.2.1139b16–17). *Episteme*, translator Martin Ostwald explains in his editorial notes to the Bobbs-Merrill edition of the *Ethics*, describes "disinterested, objective, and scientific knowledge"; the latter term poses some danger of misleading, as Aristotle's sense of scientific knowledge would, as one might reason, be rather distinct from common usage (307).[3] Techne is commonly described as an art or craft, rhetoric being a useful example of a techne, medicine another, a mastery of a musical instrument, and the shoemaker's practice, too (315). Thus, techne is concerned with production or application of knowledge, distinguishing itself by practice. Techne and episteme might, then, be understood as counterparts, one describing a form of knowledge as practice and the other a form of knowledge as theory. Phronesis is distinguished from *sophia*, both of which Ostwald notes can be translated as "wisdom" or, more specifically in Aristotle's work, "practical wisdom" (312). "Practical wisdom," Aristotle tells us, "is concerned with human affairs and with matters about which deliberation is possible" (*EN* 6.1141b10).[4] Thus, practical wisdom requires the possibility of change within what is being considered, in contrast to the unchanging universals with which *sophia* is concerned. Further, within Aristotle's framework, phronesis has moral dimensions. The moral dimension concerns "what is just, noble, and good for man"[5] and ensures that, along with our moral excellence, we "use the right means" when taking aim (6.1143b20–25, 6.1144ab5–10). Aristotle continues to explain how we might identify such a person: "The most characteristic function of a man of practical wisdom is to deliberate well" (6.1141b10). Someone who deliberates well is someone who "can aim at and hit the best thing attainable to man by action" (6.1141b10). For Aristotle, phronesis deals in particulars, unlike the universals of *sophia*, because "it is concerned with action and action has to do with particulars" (6.1141b15). Such a distinction has a clear implication for our consideration of expertise.

Aristotle makes the case plainly for us, continuing to explain that this definition of phronesis as a practical wisdom is concerned with particulars and thus actions. For Aristotle, phronesis "explains why some men who have no scientific knowledge are more adept in practical matters, especially if they have experience, than those who do have scientific knowledge" (6.1141b15). Isocrates, too, understands experience to be a crucial component of practical wisdom.[6] For a rhetorical education, experience (*empeiria*)—along with natural talent (*physis*) and knowledge of the art (*paideusis*)—is the key to turning a knack for eloquence into something more profound. Experience, along with practice, is what allows an orator to "sharpen the reasoning process and develop good insight and sound judgement" (Poulakos and Poulakos 1997, 87). Poulakos and Poulakos explicates how experience helps transform knowledge, writing that "experience wrests doxa [opinion] from tyche [luck], then, by guiding doxa toward practical wisdom, that is, toward the place where intellect and imagination, reason and emotion intersect" (88). From this, we obtain practical wisdom, which is necessary for proper deliberation.

Deliberation on difficult matters, matters of uncertainty, is where practical wisdom is required. "He who has it," Hursthouse (2006, 285) writes of phronesis, "unlike those who have not, characteristically attains 'practical truth'; that is, he gets things right in action in what we would call 'the moral sphere.'" Importantly, as Hursthouse reminds us, the dilemmas that are faced must be difficult. An older sibling stopping their younger kin from shattering someone's porcelain, even if the younger sibling will be upset having to stop, is not a decision that requires phronesis (289–91). And this example raises a critical issue we must address in terms of the intellectual virtues. The virtues may, to some degree, be natural to a person, but virtues must be habituated to be full or complete. Phronesis, and the other virtues in Aristotle's account, must be cultivated, embodied, and enacted—that is, habituated.[7] The proper and complete cultivation of phronesis produces the *phronimos*. For Aristotle, the *phronimos* is a wise, virtuous person who normally makes the correct and right decision, as this individual is in possession of phronesis, or what is often called prudence or practical wisdom. It is instructive to broaden how we might think of the qualities of the *phronimos* to help clarify the role of practical wisdom, with its moral inflection, to the question of expertise and experts. Gallagher (2019) offers a general model for understanding the *phronimos* as an "exemplar." He explains, "Exemplars are people who take action, make choices, and have a sense of proper motivation with respect to a particular disposition," adding, crucially, that the "advantage of this paradigm is its inordinate flexibility" which is rendered as individuals "exhibit[ing] a diversity of virtues while offering a normative way of acting" (140). The punchline to this is that virtue ethics provides, "counterintuitively," Gallagher notes, "a contingent normative framework" (140). This is to put forward a model rather different than one of a moral or ethical expert. Indeed, this is to acknowledge the situatedness of the expert and their insights gained through experience and immersion in a particular domain to cultivate a kind of moral framework appropriate to their expertise.

Millgram (2015) helps broaden the notion of morality as a kind of regulatory function in a specialized domain, by specialized persons. He explains in this model that "a central function of morality is regulating the generic interactions of (as it turns out) differently specialized persons, occupying different social niches" (231). Further explaining this process, Millgram writes that "niches come and go, and moreover, niches undergo internal changes, *inter alia*, to the systems of standards that govern activities within them," which is to emphasize that "whatever the content of an adequate morality turns out to be, we can expect it to be so only temporarily" (231). Such a framework does not discard more general ethical frameworks, although here I would set those aside in favor of virtue ethics' situational preceptive,[8] but it does help us understand how we might think about moral frameworks in specialized domains or niches. Further, understanding phronesis as a kind of "regulatory function" in "specialized domains" reminds us that that function is situational and, in being so, also contingent not on a universal conception of what is wise, but one accessible to those people inhabiting various specialized domains, as well as broader social milieus.[9] Indeed, our language of ethos[10]—ethos itself, eunoia, arête, and phronesis—all implicate an ethical orientation toward our audiences, and discussions of trust, too. All these matters invoke questions of the moral commitment of experts toward those audiences who might trust or distrust experts. Carr, Arthur, and Kristjánsson (2017) summarize why the idea of virtue is important in the professions. First, they note the increasing interest in virtue ethics among those studying professional ethics. Virtue ethics may be of interest to professional ethics, Carr et al. reason, because "if it is important in some professions to be *seen* to be virtuous, it is obviously no less important in all professions to *be* virtuous in some or other respects" (7). When we understand the expert or professional as a rhetorical agent, then, we must attend carefully to matters of their ethical comportment should we wish to understand or even, audaciously, advise in matters of expert engagement with their own communities of practice and with publics.

Essential to understanding the rhetorical conception of experts advanced here is that their knowledge, their practice of expertise, is situated and context-aware, but also relies on knowledge and experience to navigate with practical *nous* (intuition), *sunesis* (comprehension), and *gnōmē* (discernment). Hursthouse (2006) has examined how the *phronimos* obtains the requisite practical wisdom, which is instructive to our study of experts. Because, for Aristotle, we are born with natural virtues, we have some capacities that, if nurtured, might allow us to excel to become properly habituated into full virtue. Nurturing into such a state requires experience, and so the inexperienced and young are not likely to have what he would call full practical wisdom or phronesis because they lack the necessary experience. Although Aristotle's thinking was outside of what we now have established are studies in expertise, the prevailing sentiment in much research on expertise, too, is that experience is important. Practice-based models where numbers of years or hours are used to discuss progression toward expert status and the mastery of knowledge and skills to achieve expertise mirror the discussion of how one might achieve practical wisdom. But what experience means and how one cultivates the capacities of practical wisdom is nuanced.

Gnōmē (discernment) requires experience, and the "experience of exceptions" affords such cultivation of discernment (Hursthouse 2006, 292). This is because the inexperienced or the novice, in the terminology of expertise, is likely to apply general rules as a heuristic to a situation. While the experienced nonexpert seems to simply accrue experience but does not use it to develop a sophisticated understanding, the expert experience with exceptions or edge cases will build experience into a model to deal with more complex cases—cases requiring expert thinking. Cultivating a repository of exceptions allows our prospective *phronimos* to better understand those contextual, particular features of the situation necessary to make an appropriate judgment in that given case. It is notable that this dialectical process mirrors the conception of expertise as a dialectical process where domain knowledge is crucial but informed by particular cases and vice versa. Cultivating one's *gnōmē* will help in the development of practical *nous*. Notably, Hursthouse draws our attention to the situation itself. For to properly read and respond to a situation, one must indeed know what the situation really is about. Here *sunesis* becomes especially relevant as one must properly understand the situation and the appropriate response, and this requires careful assessment of how the situation is realized: "A 'situation' which calls for my doing something may not be facing me at all, waiting for me to read it, but rather something whose details I have to work out from what other people say about it. And until I can make a correct judgement about their accounts of the relevant matters, any practical conclusion I reached about what to do in 'this situation' would be made in the dark" (Hursthouse 2006, 293). This is to suggest that the *phronimos* is not faced with what we might think of as a Bitzerian rhetorical situation, standing before us with the exigence clearly marked by imperfection and driven by an urgency, as Bitzer suggests (1968, 7). The situation is rather more like Miller's (1992) recovery of *kairotic* moments, following Gorgias, where *kairos* serves an inventive function. The situation is not simply to be seized, but rather is an opportunity for judgment, and appropriate judgment in this moment relies on comprehension and discernment to properly adjudicate. Here, in the domain of phronesis, the situation, however, must be understood in its distinctly human terms. The activities of comprehension and discernment rely on one's capacities with respect to their fellow humans, their thoughts, their feelings, and their beliefs. The requirement to appropriately comprehend and discern what others say about a situation is where one must be vigilant to understand, Hursthouse contends.

Vigilance is important because one needs to assess whether the situation is indeed accurately conceived and described, but also that there is some dilemma, some imperfection Bitzer might say, that must be resolved in the first place (Hursthouse 2006, 294–95). Sherman likewise argues that "preliminary to deciding *how* to act, one must acknowledge that the situation requires action," adding that a reading of the situation "is informed by the ethical considerations expressive of the agent's virtue. Perception is thus informed by the virtues" (1989, 29, emphasis in the original). Thus, it is the capacities of the *phronimos* that allows for the appropriate deliberation on the situation itself as well as the available means to respond to a given situation. Illuminating to expertise and experts, this argument provides nuance to claims, such as Engeström's (2018, 111), that "experts produce decisions," and that such decisions "are the main products of expert works." Engeström illuminates how moral knowledge is related through his activity theory approach. He explains that decisions are not strictly produced by a single, individual expert but rather "are typically steps in a temporally distributed chain of interconnected events," adding they are not "purely technical" and have "moral and ideological underpinnings with regard to responsibility and power" (111).

These capacities, in addition to moral knowledge, draw on those technical or practical kinds of knowledge that are cultivated in part through experience, including that of exceptions (*gnōmē*), as well as through reflection and learning on those experiences. Cultivating the capacities to become a *phronimos* overlap, I contend, with the capacities required to become an expert. Where the two ideas are distinct is in scope, as the *phronimos* is a person who has moral knowledge and practical wisdom that might advise others on how to live a good life, and the expert, in the model I put forth, is rather a person who has moral knowledge and practical wisdom and applies it in a specialized domain (specifically, when faced with an audience not privy to the inner workings of that specialized domain, the expert must offer not only technical but also morally grounded advice). The expert's phronesis is, thus, not a specialized form per se, but it is restricted. We might consider Aristotle's fondness[11] for comparing the practical knowledge phronesis details with the techne of a physician. A physician has episteme and techne,[12] the former normally cultivated through an educational program providing foundational knowledge on the human body and its systems, and the latter cultivated through residency and, later, daily work in a medical setting. However, physicians must also have moral knowledge as their daily activities require informing patients of risks they may face with a particular procedure, for example, or making difficult care decisions at the end of life. We might also recall those cases of pharmacists who refuse emergency contraceptives to patients on the basis of their own religious beliefs. In the former case, the moral knowledge that the expert, the physician, can bring to the situation is specific to the particular patient interaction born not of individual moral belief, but rather professional experience, social agreements, and phronesis. In this case we might ask the physician to weigh the relative risks, based on experience, for care at the end of life relative to less invasive treatment and a more palliative approach. Whatever the treatment plan will be must value the patient's perspective, as a basis of professional practice, as a kind of specialized domain for moral decision-making, as well as a social agreement about care. Conversely, our other case finds that the pharmacist is rather operating in a prescriptive moralistic fashion, divorced from the situational elements we wish to consider, and also divorced from professional duty of care and social agreements about the expert's role. In the latter case, such pharmacists cannot be experts not because they are unfamiliar with the technical aspects of preparing or delivering prescriptions to patients, but on the moral grounds of failing to adhere to professional values comporting experts toward their audience, values that in turn shape the anticipated rhetorical situation to which patients imagine they are about to engage. In each case, we also see that certain social agreements, too, participate in how prudential meaning is constituted. Indeed, to understand fully that phronesis is not only individual reasoning but also a manner by which the "nodular self is itself fully activated only by fulfilling its centering function in respect to a potential group," one must also consider the audience and broader ecologies within which prudent thinking occurs (Hariman 2003, 305).

Expert Virtue

It is useful to first imagine "the expert" as an entity constituted by individuals, their social and professional networks, and, crucially, their ongoing habituating to act expertly—in the vein of Aristotelian thinking, to continually become a *phronimos*.[13] Further to Aristotle and conceptions of phronesis, Ding (2007), in a study of the *Analects*, demonstrates a kind of rhetoric concerned with habituation rather than persuasion alone. "Confucius," Ding writes, "is mostly concerned with the means of *influencing people's behavior and moving them to action* through *exemplary conduct* rather than through speeches" (150, emphasis in the original). Further to this point, Ding argues, Confucius "considers virtue as a lifelong pursuit in and of itself because it is the only path to becoming a persuasive rhetor" (154). Mailloux (2004, 458) offers the common conception of the *phronimos* among commentators on Aristotle, writing that it "involves deliberations over what is the good for humans in particular situations. It determines right action in such situations." To constitute the *phronimos*, the person in question "possesses experiential knowledge of all that is useful in judging and achieving the good in life" (458). Although there is something of a global orientation in this formulation, the *phronimos's* way of being is always locally grounded, and "rhetors must have *phronetic* insight into concrete situations to determine the *kairotic* moment for the most effective speaking" (463). For Aristotelian conceptions of our intellectual virtues, phronesis is crucial to one's action in the world, allowing one to have knowledge of the correct pathway to making a decision, although not necessarily the correct decision itself.[14] Phronesis, for Aristotle, was also a deeply moral orientation toward rhetorical activities. Mailloux notes that phronesis is foundational to Aristotle's conception of rhetoric itself as he "builds the *phronetic* power of discernment into his influential definition of rhetoric as the ability, in each particular case, to see the available means of persuasion" (458; cf. *Rhet.* 1.2.1355b). Thus, of experts with phronesis, we might say they can be wrong, but have access to richer knowledge-building resources, relative to nonexperts, to move them through a problem requiring expertise toward viable solutions.

Self (1979, 135) reminds us that phronesis "involves an inherent social orientation and responsibility." Further, phronesis "is not simply self-serving or egotistical" but rather "concerns itself with one's self, one's family, and the state because the individual's welfare is bound up with that of others" (142). That is, the person with phronesis must embody arête (moral excellence) and act for the benefit of not only oneself but also one's audience. For experts, this might be a matter of an internal audience of other experts or a more public audience. In the former case, the relationship with the audience of other experts may also be constitutive of the expert's moral comportment, as a constitution of the expert character. For Aristotle, the *phronimos* requires a community of *phronimoi*[15] who serve to model the requisite habituated virtues, and we can extend this idea to the importance of phronesis in expert communities (Sim 2018, 195). Importantly, however, we are reminded that ascribing a moral component to the expert, in the sense developed here, does not position that individual as a moral expert.[16] Rather, through this configuration of experts, further to their social and professional networks, a crucial socio-rhetorical constitution of experts is in their audience. Experts exist quite relationally, in changing configurations—not unlike the configurations required of Aristotle's *phronimos*—and are fundamentally directed through a phronetic orientation to act in response to a particular situation with their specialized available means.

Underlying experts is their expertise, which can be of multiple natures. Normally we might say expertise is constituted, in part, as techne, a craft where the products might move from less to more expertly crafted. We would also normally say that expertise requires episteme, a universal, unchanging theoretical knowledge often characterized by those insights we today ascribe to scientific knowledge.[17] Along with these qualities, good moral judgment is required and is directed through the enactment of expert status through phronesis. Put simply, in addition to theoretical knowledge or understanding and applied skills or craft, experts must also have a kind of practical knowledge, a practical wisdom, of how to respond to a particular situation—and, importantly, this knowledge is moral in nature.

The mere capacities of technical knowledge without the capacities for prudential thinking—without phronesis—is not meritorious of expert status. A rhetorical perspective[18] illustrates this in that the activities that render experts or expertise meaningful occur in moments of collaboration or argument among experts or deliberation with publics. Questions of trust point us to this conclusion. Trust is a concept that reminds us of the rhetorical nature of expertise. For experts to enact their expertise in a manner that will foster trust, they are engaging not only in a techne but in phronetic activities to understand situations (i.e., survey the available means with which they might respond and deliberate and judge a course of action). Expertise, too, requires moral knowledge to functionally operate because expertise is embedded within a community of practice where the values and norms of the community shape practice. Inevitably, some might be skeptical about the requirement of moral knowledge in the role of expertise. Consider, however, when we think of the ways in which experts and expertise are challenged: claims about scientists biased by career advancement, doctors under the thumb of big pharma, and so on. Or, from another vantage, consider the ways in which experts and expertise are deceptively mimed: bogus medical staff, profit-driven "wellness," investment salespersons dressed up as financial advisers, and so on. The moral nature, in addition to technical or scientific knowledge, of experts becomes quite apparent. Drawing on this line of thinking from antiquity, the rhetorical formulation of experts and expertise I wish to advance here examines the crucial role of phronesis. For experts to deliberate on their expertise, they cannot rely on technical knowledge alone, for they do not operate purely in the realm of the technical. Experts have technical knowledge, for which we value their specialized knowledge, but to apply their technical knowledge in a phronetic manner is that for which they are awarded the title of expert.

The essential argument going forward is this: expert status and expertise—in addition to and crucially being a matter of knowing-that (episteme), knowing-how (techne), and attribution by an audience—is a morally grounded activity of knowing-why (phronesis). In this sense, I mean morally grounded as a pragmatic rhetorical activity. Indeed, I am not a moral expert, and intend to make no such claim. "Que sais-je?" after all. Experts broadly are not moral experts, either, and I do not make that claim. But experts are required to make moral judgments in the regular course of their being expert. Here the moral dimension of experts and enactment of their expertise is discussed to provide a more comprehensive understanding of how experts and expertise function rhetorically. Plainly, to understand experts, their expert capacities, their expertise, their expert status, and their engagement with other experts and publics, a rhetorical account must attend to the moral qualities of expert activities in practical reasoning. Moral in this sense calls to deliberative capacities an expert uses to respond to particular situations. When an expert is called on to respond to a particular situation, the situation normally not only requires technical knowing-that or knowing-how,[19] but also must be infused with prudence. Underlying the rationale for this model is an alignment of moral knowledge (through phronesis) with expert knowledge as both kinds of practical knowledge.

The idea of expertise advanced here, then, is not tidy. Instead, expertise is understood in its relation to expert status, and vice versa, and both in pluralities. Practically, this means that expertise is understood to be relational as its enactment is contingent on expert status with respect to an audience. Further, expertise is constituted by its social cultivation among a community of experts. Finally, expertise and expert status are required in those situations that, as Aristotle might say, can be other than they are. Situations to which a response might change their outcome are of interest here, in opposition to unchanging universal situations to which no human response might recognizably alter the result.

Social Theories of Expertise

When we speak of expertise, we in fact must speak of multiple forms of the phenomenon. Not only is expertise multifaceted, it is multivarious. Among those who have provided some account of the forms of expertise, the work of Collins and Evans (2007) is among the most widely adopted.[20] In their model of expertise, they provide both a clear rationale for discarding simplistic divisions between experts and publics and a model to understand expertise as it exists in many forms. In such a model, publics, too, may have forms of expertise that contribute to specialist discourses. Collins and Evans offer a "periodic table of expertise" to help disentangle forms of expertise, offering a kind of gradient from what they call "ubiquitous" expertises to "specialist" expertises.[21] Ubiquitous expertise, as the construct suggests, comprises those kinds of expertise that are somewhat common among a population. One's own language acquisition and performance is such an example, they tell us, saying that we must have some abilities or skills to become proficient, but we would not typically notice such proficiency as an expertise. Such ubiquitous expertise is to be distinguished from any simple task that almost anybody could perform. Lying in bed all day, they explain, is an example of something anybody can do, but one certainly does not become an expert in lying in bed all day (they do note where the term expert might be used in a sort of disdainful irony) (55). Specialist expertises appear in two varieties with some speciation within them. The first is "ubiquitous tacit knowledge," which is inclusive of "beer-mat knowledge" (gained through some short explanation of a concept; Collins, for example, cites learning how a hologram works from reading such a description off a literal beer-mat, a British term for a drink coaster used in bars), "popular understanding" (a kind of knowledge more extensive than beer-mat in that it requires engagement with, for instance, popular science books), and "primary source knowledge" (a form of knowledge gained by reading primary literature but not being immersed within a scientific field such that understanding the debates in the field or level of certainty in the science is challenging) (18–22).

The second variety is "specialist tacit knowledge," which includes "interactional expertise" (knowledge of a specialized language; for example, knowledge of terms used in rhetorical theory) and "contributory expertise" (knowledge that allows the doing of research; for example, the ability to perform rhetorical criticism or, in Collins's example, build machines to study gravitational waves) (23–35). Interactional expertise is a kind of expertise developed by immersing oneself in a research discourse, learning the language of the community of practice, and being able to communicate that information to others who are unfamiliar with said research area (generally the public, but perhaps also others involved in research processes who do not understand the science but are required to interact with the field). We can imagine an interactional expert, for example, working with engineers on a space program, who helps communicate their technical problems to a broader staff involved in planning missions, funding, or reporting to governmental officials. Interactional experts are nearly full-fledged experts in every respect except that they cannot themselves *do* the science (or otherwise contribute to the area in which they are expert). Contributory experts are those can *do* the science, and we might typically associate them with the more general concept of an expert.[22]

As an expert in psychology, Siobhan specializes in "children's well-being in adversity." Becoming expert in this area, for Siobhan, primarily occurs through readings in the field, and with some expertise gained by working with other experts.

Indeed, the latter case of working with others in an academic field is well supported by Collins and Evans's model of expertise, which would also support reading in the field as primary source knowledge to become expert. Training students provides important insights into Siobhan's understanding of expertise. When training students, Siobhan uses a strategy of "presenting conflicting evidence" in an effort to "try to get them to think critically on that evidence." Although this functions differently for undergraduates (who may be presented with conflicting information in a lecture) and doctoral students (who may be expected to perform such comparative work in the course of their studies), the purpose is to develop critical thinking capacities. Achieving these critical thinking capacities requires becoming encultured within a particular discourse community, learning a particular form of critical thinking within a disciplinary framework. When Siobhan cites reading papers in their field as crucial to the cultivation of expertise, it would be easy to categorize that simply as acquiring knowledge of the field, knowledge as episteme. However, when steeping oneself in the literature of the field, other forms of knowledge are cultivated, such as the ability to speak the language of the discipline in written form. Reading papers in a field, however, is not enough to become fully immersed within the community. For Siobhan, it is also important to work with other experts. In addition to these capacities or knowledges that help cultivate expertise, Siobhan also acknowledges the communicative aspects of expert status: "How I'm perceived as an expert, I don't think that the reading helps. It's done in the publications." Distinguishing between one's own reading and one's publication record, however, may be misleading. Indeed, the relationship between what one reads in a field and the related kind of domain knowledge they acquire is not merely a record of those articles in their head, but rather a larger source of (ideally) integrated knowledge produced by a kind of higher or deeper learning. There are, however, aspects of expertise and expert status that do seem to be less principled on one's treasure house of knowledge and experience.

Quinn, an expert in communication studies, on the matter of becoming expert, explains, "I think of it as an acculturation process. It's a combination of disciplinary training and engaged research . . . if we're talking about academic disciplines [expertise not being limited to this domain, but as an example] then the marker of expertise is the ability to publish in the academic journal. And so that is the highest level of expertise [in an academic field]." Part of achieving these levels of expertise requires, in Quinn's training model, acculturation through apprenticeship. Caden further illuminates how we might imagine this process of acculturation at the level of the discipline, explaining that, for them to become expert, "conferences were a big one." Caden tells us about their acculturation through conference-going, explaining that they attended a larger conference "three or four times in my graduate studies." Conferences help acculturate one to the discipline, Caden believes, in part because "you're just bombarded from all sides with the cutting-edge research in your field." In addition to this kind of professional acculturation, Caden also explains that university-level courses function to provide content knowledge and also, at the doctoral level, to administer comprehensive exams to situate one in a field. Coursework, Caden explains, was a "big one" for providing broad knowledge of the field, detailing their interests: "I took a few special topics in advanced perception, or kind of advanced visual perception." Later, Caden tells us how during their doctoral work comprehensive exams were important in gaining field-specific expertise, noting that exams not only train individuals in their specific subdiscipline but "[push] you into slightly separated fields than the one that you're working in." Pushing one into a slightly different field is "useful because I used to think that I knew a lot about those fields, and then I started reading a lot more in them and realized that I did not know as much as I thought [*laughter*]." A key takeaway from such a lesson Caden provides that links this process of acculturation to a discipline is explicitly ethical: exams are "a good practice for getting some humility." Here the link to humility as a central idea for being capable of understanding the limits of one's knowledge connects directly to the capacity to become expert. One must be aware of these limitations to determine how to assess not merely routine cases, but the kinds of edge cases we would expect an expert to be able to deliberate and act on. Maintaining this quality of humility throughout one's career, not only at the beginning is a central requirement for the expert, and is a quality supported by the cultivation of one's phronetic capacities.

Expert Individuals and Communities of Experts

Attention to the expert as an *individual* provides an identifiable nexus for a multitude of conditions and practices required to sanction knowledge claims within a particular field. However, expertise cannot be theorized solely in individual terms. Rather, expertise and those whom we call experts are part of a more complex system than individual notions of expertise or experts allow. Jasanoff (1998) demonstrates how expertise and expert status is communally negotiated in a study of scientific expertise in legal settings. Her analysis examines claims to scientific expertise in disputes in the United States about the health effects of silicone breast implants. Using the model of an "expert games," Jasanoff shows that credibility as an expert is gained by appeals to objectivity or professional status, and credibility is lost when appeals suggest an expert is subject to or may have biases (103). In legal disputes she argues that judges have considerable influence over the parameters for expert claims. Jasanoff argues that this negotiation of expert status and expertise can bring forward negotiations around such status that might otherwise seem given, including expert status and the role of judges (104). Expert status is important to distinguish from expertise if we wish to understand the relationship between attributional qualities of expertise and those qualities that are embodied by an individual (fully granting, still, the important situatedness of that individual in a community and relationally to those other experts or nonexperts who rely on said individual's expertise).

"By taking the job" is how Roshan jokingly responds to the question "How do you think you became an expert in this area?" As an expert in forensic DNA analysis, the fields of forensic science and forensic DNA are not regulated in the countries where Roshan works. Raising an important and often overlooked aspect of expertise, Roshan explains that one measure is the "legal recognition of someone being an expert in court systems." This criterion has various definitions and norms across court districts, but most basically it means "knowledge a layperson wouldn't know about a topic," meaning that, in some instances, "having a college degree could technically qualify you to be an expert in a field." Emerging areas of research are interesting cases to explore questions of expertise and expert status because often the debate about who an expert is and how to become one are more transparent than in established fields by virtue of their ongoing negotiations. Roshan completed undergraduate studies in computer science and gained inter- and multidisciplinary capacities through working in the field of forensic science. Computer science, however, also offers an interesting case for questions of standardization, regulation, and, indeed, expertise. Noting the interviewer's discipline of English, Roshan explains, "The study of English has been around for so long. But mechanical engineering, physics, chemistry—these are hundreds of years old. Maybe two hundred years old, three hundred years old. Since the scientific revolution. But if you're familiar with professional engineering qualifications, civil engineers building bridges, laying railroad track, they have to be qualified by independently accredited bodies."

Differences in disciplinary history, in the development and socialization of would-be experts based on those histories, is an important distinction to make when we talk about expertise. Not all disciplines, including those with notable social status, can be compared quantitatively: "So they just created that test for software engineers. And I think the first year they offered it sixty people took the test in [country name removed]"; yet, Roshan continues, "there's a million people, according to the federal government, . . . whose primary job is software development. And sixty of them took an exam that will allow them to call themselves professional software engineers or to be fully accredited," adding that that creates "a huge unregulated disparity." Roshan, perhaps given this experience working in those less defined fields, offers insightful and nuanced ideas about training. Training, because the field of forensic science is so multidisciplinary, requires a tailored approach to individual students' backgrounds: "A student who's perhaps in their third year of an undergraduate biology program is going to have a different understanding than a student with equivalent experience in an engineering discipline, a computing discipline." However, because the field is new, a general primer by way of instructional videos helps introduce trainees to this new field of forensic DNA analysis. Although each student is a unique case, and "expertise in another field is going to influence how fast or slow" students are able to master certain concepts, Roshan tells us that novices will need to learn the methods of statistical analysis and data analysis required of the multidisciplinary problem space to become expert in this domain. "Like any field," Roshan further elaborates on the path toward expertise, "critical thinking is the single most important thing and if you can't get those first principles into their brain so that they can work from that foundation, then you're programming automatons as opposed to actual human experts." It is this idea of critical thinking, often the skill we in the humanities—along with those in the social sciences, physics, math, and most other disciplines—promise, that seems central to becoming expert. Knowledge must be situated not only in one's own schema, but within the broader culture of expertise within which one attempts to participate.

Engeström (2018) further advances a definition that moves understanding of expertise from an individual into a collective engagement. Offering three propositions to do so, he expands how expertise can be formulated as a relational concept. In his account of expertise, Engeström challenges common units of analysis by moving the locus of expertise from the individual to the activity system.[23] Moving from the individual to the activity system challenges a great many models of expertise, but also provides a manner in which to seriously engage relational, social, or attributional models of expertise. Indeed, Gobet (2016, 250) reminds us that an understudied area of expertise has been among teams. Focusing solely on the individual, Engeström argues, does not provide a full account of expertise. Instead, a richer sense of expertise can be developed through an investigation of collaborations and, crucially, activity systems. Engeström's second move to account for the composition of expertise notes that expertise is not simply specialization. Such an understanding creates and once again moves expertise from the individual to team- or collective-based understanding[24] of actors operating within object-oriented activity systems, where it is not only the individuals working together but the broader objects, artifacts, networks, rules, and social dynamics that shape expertise.

With a bachelor's degree in bioengineering, specializing in synthetic biology and related areas, Rory had considerable interdisciplinary training and important insights into this question of teams and collaboration. Now working as a scientist in citizen science projects, Rory conceived of citizen science as "a citizen's participation in biotechnology and in science," and of "science as a formalized kind of [work]: have a question, have a hypothesis, experiment, and iterate. That formal process, doing it with yourself or with a group. That's what I define citizen science to be. When it becomes completely detached from academic background." Although Rory has some sense of what is meant by the enterprise, they do not identify as an expert. Rather, Rory does acknowledge they have both formal and self-taught informal training in citizen science projects related to biotechnology. Through a "combination of classwork," a closely related internship, extracurricular activities, their personal "interest," and "taking [personal] time to kind of get to know the field doing my own research," Rory was able to gain a considerable range of training and experience that has put them on a path toward becoming an expert in their field. Rory's reluctance to affix an expert label on themselves stems from a vision of expertise as a rather more social form of knowing. Rory explains, "I think expertise becomes a very blurred line. . . . I believe most people who are interdisciplinary think that there is no such thing as a person that is an expert in everything that they're doing for their interdisciplinary lab or their interdisciplinary work. . . . If you're part of the center that's trying to hold this together, expertise doesn't rest on one person. It rests on a body of people. So maybe it's fairer to say this group or this entity or this collection of scientists has the expertise. Rather than this person or that person has expertise."

The Challenges of Assessing Expertise

Across the multidisciplinary literature on expertise, repeatedly we learn that expert performance, expert knowledge and skill, and assessment of expert status depend on the context or situation under which expertise was cultivated and to which expertise responds as an answer, or at least as some form of intervention. A rhetorical perspective might here also emphasize how such contexts and situations are crafted and how such crafting embeds norms and values. Indeed, attending to the ethical choices embedded in these contexts illuminates strategies for assessing expertise. Indeed, ethical norms importantly shape expert engagements among experts, and among experts and nonexperts, too.

Divorcing expertise from a mode of ethical engagement renders expertise something else entirely. Chess players, for example, are constrained by rules of fair play. The International Chess Federation's (2020, 08-1-1.1) FIDE Code of Ethics states that players must observe the rules because they are important to "fair play and good sportsmanship." Even in this game with satisfyingly complicated measures of success through Elo ratings, the ethical engagements of players remain beholden to one's good reasoning. Consider the following excerpt from the Code of Ethics: "It is impossible to define exactly and in all circumstances the standard of conduct expected from all parties involved in FIDE tournaments and events, or to list all sets which would amount to a breach of the Code of Ethics and lead to disciplinary sanctions. In most cases common sense will tell the participants the standards of behavior that are required" (08-1-1.2).

Thus, even in this game where the individual psychological attributes leading to success are the typical measure of expertise, we see that ethical and moral virtues are central to expert performance. Phronesis is crucial here for judgments, as "common sense," regarding how to assess and perform in an ethical or moral manner in "all circumstances," not simply in a rulebook fashion. What occurs when one violates ethical standards or fails to inhabit virtues is that such actions diminish one's expert status. Consider violating rules in a game of chess, where it would seem the rationale might be to win a game against someone you otherwise could not beat. The recognized act of cheating, more so than losing, marks a rhetorical loss of expert status. Does the person really lose their expertise, though? Given that expertise is not entirely based on an attribution, an instance of losing credibility, such as in a singular case of cheating, does not per se result in the loss of expertise. Perhaps such a singular case could result in the loss of expertise if one were not only disqualified for a match but more permanently barred from

competition. Transgressions that violate the ethical norms of one's expert community may lead to a loss of expert status and, by the loss of status and engagement with other experts, loss of expertise itself. Provided experts can still perform
among other experts, then the danger comes in the form of habituating themselves
to cheating, lessening their specialist knowledge and skills, and the critical deliberative capacities that allowed them to enact their expertise among other experts.
Indeed, expertise in the rhetorical formulation advanced here is one of an unfolding rather than a permanent status. That is, a status that can be gained, lost, or
perhaps recovered, but is always in need of tending.

Another matter complicating how we estimate the capacities of experts is
where we draw boundaries around their expert knowledge. This raises important
distinctions between matters of specialization and profession. Chess players are
a rather straightforward example of experts with measurable status through Elo
ratings. Most experts, however, operate in less measurable circumstances. "I lucked
into it," Avery explains of becoming an expert in geometrics, an area of applied
mathematics. Meeting other researchers in this area and then spending considerable time working with them, along with continued studying and "curiosity,"
were critical to becoming expert. For Avery there was an affective element: "It's
what I love to do, so I just sort of kept reading." One might think expertise in
mathematics is a straightforward enough problem that it is easy to measure, but
Avery works across disciplines, with several significant collaborations with
medical researchers and, later, with an engineering researcher on robotics. When
we imagine these applications of mathematics, it is not difficult to see how
quickly the expertise necessary to address a particular problem or problem space
will become increasingly complicated. Mathematics instruction itself, to return
to a more restricted case (by discipline) offers a problem that provides some
insight into the challenges of becoming expert. Avery explains a major challenge
in mathematics instruction: "Once you see an answer, it makes sense." Even
once this happens, and the answer makes sense, this does not mean one is able
to derive the answer. This, then, poses challenges to having students work
through the problem itself. "You need to sit and not understand what's going
on with this homework problem for a little bit," Avery explains to students.
Sitting with the problem, even when one does not understand, is crucial to development of expertise in Avery's mind. "I think that that's really important," Avery
argues, "and that's sort of how I gain expertise." For Avery the process goes
something like this: "That doesn't make sense. Why doesn't this make sense?"
When teaching the subject, Avery provides moments where students must

themselves work through a problem during the lecture. We might, returning to the question of assessing expertise, ask how we could measure the expertise of any given subject. In an academic setting, tests and exams are one measure, but even those that would provide partial marks for getting the solution nearly to a correct answer are marred by other factors, such as how well certain students can take tests. When we then move these individuals into a space where they are applying their knowledge to problems, for instance with medical experts, we further lose the constraints that allow for easy assessment of expertise.

The role of moral knowledge in this process may not be clear, but it is indeed present. Although moral knowledge is not emphasized in Avery's account of expertise, we could easily find it in discussions of exams and academic honesty, collaboration with those working in medical fields, engineering ethics, and so on. Absence of moral knowledge as an explicit conversation is indeed the key erasure I wish to address in the theorizing of expertise. Ultimately, the call here is to recognize that experts must act with moral knowledge, and whether we acknowledge such a comportment is rather a matter of how we rhetorically understand the role of experts. Whatever the case, such knowledge is foundational to an expert's very capacity to cultivate and then use their expertise. A more rhetorically sophisticated account of experts centers the key form of practical, moral knowledge alongside knowing-that and knowing-how to emphasize knowing-why.

2

Expertise in Rhetorics of Science

In rhetorical studies of science[1] Lyne and Howe (1990) offer a conception of how one becomes an expert in their investigation of E. O. Wilson's effectiveness[2] in advocating for sociobiology across his own field of specialty, to other sciences, and then across fields of the social sciences and humanities.[3] They write that "expertise presumes a functional inequality of knowledge among different groups," and that an "expert is tapped when non-experts need access to a special knowledge they do not have" (135). However, according to them, expertise is not limited to an expert-public relationship; instead, "expertise functions rhetorically from the moment expert communication begins *within* science" (135, emphasis in the original). In their analysis, Lyne and Howe follow how Wilson can move from his specialty of entomology into sociobiology. Charting this progression from disciplinary to interdisciplinary to extradisciplinary audiences—first authoring *The Insect Societies* (1971), then *Sociobiology: The New Synthesis* (1975) and *On Human Nature* (1978)—Lyne and Howe demonstrate that Wilson can adapt his arguments for sociobiology to multiple audiences to highly persuasive effect. Although Wilson ultimately claims to offer a modern, rational worldview that might supplement or supplant old fashioned humanities, he strongly relies on a rhetoric "disguising itself as the affectively neutral entailment of the scientific theory" (Lyne and Howe 1990, 147). To accomplish such a stance, Wilson's program of sociobiology relies on expertise operating as a kind of public performance of scientific ethos, but not scientific argument.[4] Complicating this matter, as Lyne and Howe argue, is that expertise is not simply specialization and if we take it to be a "rhetorical construction, applied within social formations," then we cannot simply use "epistemological criteria" as the standard of measure (148). Indeed, following Lyne and Howe's early work and the growth of rhetorical studies in science that occurred in the late 1980s and early 1990s, expertise has since generated much discussion among rhetoricians interested in science. Notably, Hartelius (2011, 1) also centers ethos at the core of expertise when she writes, "Expertise is not simply about one person's skills being different from another's. It is also grounded in a fierce struggle over ownership and legitimacy. To be an expert is to claim a piece of the world." Because judgment about one's expert status is challenging for those who exist outside a narrowly defined group capable of assessing a full range of expert competencies, a great deal of expertise is adjudicated not by technical knowledge, but by character.

Ethos and Expertise

In rhetoric, ethos offers great insight into how a speaker cultivates their credibility through character-based appeals, demonstrates their expertise, and gains the trust of an audience. For Aristotle, ethos was the "default appeal" (Miller 2003), the appeal to which we turn when logos or pathos are not perhaps the best available means one has to persuade their audience. Ethos itself comprises three components in the Aristotelian tradition:[5] *arête* (good moral values or virtue), *eunoia* (goodwill toward the audience), and *phronesis* (practical wisdom). Ethos and ethics, in the enactment of knowledge, mark how one chooses appropriately from the available means and then applies judicious reasoning to the selection of an appropriate repertoire of responses. "Only practical wisdom," Gadamer (1997, 57) writes, "is capable of employing science, like all human capacities, in a responsible way." In this light, ethos remains an important aspect of how one constructs credibility, but it also shows us the depth of this appeal. Ethos is not superficial; rather, it is an enactment of a particular ethic.[6]

Hartelius (2011, 3) further locates the negotiation of expert status[7] within the rhetorical situation to "theorize the rhetorical strategies that experts of various specialties employ to compete for authority and legitimacy." Here the rhetorical situation is rather focused on the attribution of expert status, rather than, as above with respect to Hursthouse's (2006) account of the *phronimos*, the expert's broader assessment of the rhetorical situation, a kind of prudence (Hartelius 2011, 170–71). Hartelius details how expertise is sanctioned and why expert status is central to contemporary conditions. Experts, she argues, are necessary as everyone's daily lives increase in complexity. In her account, Hartelius demonstrates that rhetorical strategies are deployed to reaffirm distinctions between experts and publics. However, she further complicates the expert/public division by exploring how both politicians and immigration activists deploy political expertise to assert political influence, demonstrating how such divisions are not confidently maintained.

Experts must have some domain of knowledge that they claim specialized familiarity with, but for public audiences, attribution is a necessary condition for one to be sanctioned as an expert. "From a political standpoint," Hartelius writes, "being an expert depends both on possessing a particular kind of knowledge and on other people's recognition that we possess it" (6). As an example, Hartelius argues political experts rely on rhetoric because "one can function as a political expert only if the public accepts one's claims to possess certain capabilities and virtues that are superior to others" (38). "This rhetorical gesture," she concludes, "is the primary mechanism for the delegation of political influence" (39). Although rhetorical definitions of expertise normally offer a social orientation—where the expert can only "enjoy expert status to the extent that they can motivate an audience to assent"—Hartelius reminds us, in Aristotelian terms, that expertise is comprised by artistic and inartistic proofs (9). Consider, she suggests, the lawyer or economist who holds not only a degree but also a title, and "substantive knowledge in a socially designed area whether or not the neighbors know it" (9). Artistic appeals are also necessary in expert-expert interactions. That is, other experts' statuses are similarly contingent on others' recognition of one's expertise. Even in the most technical of fields, one's ability to participate is dependent on not only specialized knowledge but also an ability to participate in the technical community. Consider, for example, research scientists who have significant technical expertise in the sense they may be able to perform impressive experimental work, but perhaps have yet to master discursive norms of the discipline to accomplish having their research published in a prestigious journal. Although the scientific experiment itself may be sound, it is unlikely that without the appropriate rhetorical craft adhering to disciplinary norms that publication of results will easily follow. Such junior scientists would set about learning additional scientific-discursive skills to use alongside their more theoretical or methodological expertise to become persuasive in their claims as experts to other scientists. Attribution of expert status matters, whether by a broad public as in Hartelius's political example, or by a specialized community, as in the case of research scientists.

We can identify how such attributions become illuminating in cases of citizen science, and how there may be a disconnect between what someone knows and their status as expert. With an undergraduate degree in the humanities, Sean, a citizen scientist, provides another interesting perspective that crosses the two cultures.[8] Sean founded a nonprofit citizen science project, which involves training citizen scientists who previously had no background in science. Although a humanities degree would seem to suggest little training in the sciences, through professional courses in the sciences Sean was able to obtain skills and even certification in marine life identification and data collection. Recertification, further, takes place every year. On the question of expert status, Sean explains that the "word 'expert' is always kind of a hot-button word." Sean explains they have worked in a specialized areas as a citizen scientist and gained important knowledge and skills, but also says the principal investigators (PIs) "are the true experts." On becoming an expert, Sean cites the formal training programs in which they participated, first, and second, the benefit of experience, saying that it was a matter of "having done it for twenty years. You just gain experience in terms of marine life identification protocols and things like that." Experience is emphasized in addition to knowledge as formal training, but so too is practice within a scientific discourse community:

> I have published three peer-reviewed articles in scientific journals within my field, of course, which is marine citizen science. And so that represents a real sea change, I would say, because ten years ago, you couldn't get published if you didn't have a PhD. But now since the scientists are publishing peer review, which are then being looked—those articles are being looked at by credential scientists and now citizen science—if it isn't being accepted 100 percent, we're getting up into the ninetieth percentile here where scientists are being forced to recognize that the data we collect is just as good as data collected by graduate students and scientists. And in fact, there have been actual studies on this subject of, "Is the data collected by citizen scientists as good as the data collected by scientists?" and that's actually a paper that was published. And the verdict was, yes, for the most part, the data we collect under proper scientific supervision is just as good as data collected by graduate students and scientists themselves.

However, Sean acknowledges that there is "sometimes justifiable skepticism on the part of credentialed scientists toward citizen science." Sean then notes it can be helpful for citizen scientists to use their network to connect with professional scientists, to have them "looking over your shoulder, looking at the data, making sure you're following scientific protocols because then if you do that and you want to publish in peer review and somebody says, 'Look, dude, you've got a degree in [the humanities]. What do you know?' And I'll say, 'I've got a PI looking over my shoulder to make sure I follow proper scientific protocol.'" Embedding in a network helps establish credibility, as we have seen, particularly in attributional frameworks. However, we can see this is also a strategy required to have one's work fairly evaluated in the case of citizen science. Such evaluation can demonstrate that the citizen scientists are participating within the epistemic framework of science as well as the social spheres; in the latter case, this includes the kind of sanctioning that is provided by not only working within a particular epistemic or theoretical framework but also operating within the systems of credibility.

Crafting Ethos Outside the Professions

Archer (2012) examines the Oregon Citizen's Initiative Review (CIR), an effort to involve citizens in deliberations about proposed legislation, and argues that although experts emphasize ethos, citizens focus on claims and their support (that is, the logos) of appeals. In Archer's analysis, experts attempted to construct their ethos, attending to all matters of phronesis, arête, and eunoia in doing so. As well, citizens commented on ethos when the goodwill or virtue of the speaker was doubted, but most general citizens were rather preoccupied with technical matters. Indeed, the ethos and expertise of speakers, Archer explains, was often simply reduced to their profession when citizens spoke of their testimony. Indeed, the transcripts of this testimony, Archer argues, "reveal citizens capable of engaging in a dialogue with experts, understanding technical information, and asking for clarification when needed" (59). Thus, Archer suggests a more complex account of public understanding of science and technical information, one that implicates that logos is needed. Indeed, it is essential not to discount the capability of informed citizens to seriously engage in complex subject matter, and there have been numerous cases of citizens becoming deeply invested in even the most complex technoscientific problems for which a community must deliberate and act.[9] How and when citizens, nonexperts, or experts outside sanctioned spheres of discourse can act have been widely considered in rhetorical studies of science and allied scholarship in science studies. When experts are sanctioned by organizations such as universities, a good deal of ethos is built into their appeal through their degrees, institutional affiliation, the professional networks and bodies within which they are embedded, and so on.

Ethos is often combined with other appeals to establish expert status. Although ethos provides us with a procedural understanding of expertise through phronesis, there are certain forms of knowledge required to execute such reasoning. A common conception of experts would suggest they have prodigious knowledge of their specialty or field. Normally the form of knowledge experts are understood to possess is of a formal or theoretical type, obtained through study and mastery of the subject. Logos is one form that helps us understand further the relationship between knowledge, episteme, and reasoning. But pathos, too, is implicated in our reasoning as a vital function in motivation and decision-making.[10] All these forms of knowledge are central to the enterprise of expertise. Theoretical physicists must possess a great amount of formal knowledge, for instance, to contribute to research in that field. Their formal knowledge—attained through years of study in coursework, later during their doctoral and postdoctoral studies, and continued through their research—provides the basis for their claims to expertise, on the surface. However, knowledge is more varied than formal or theoretical knowledge alone, and considerable treatment has been given to the role of knowledge in the cultivation of expertise. Physicists, for instance, will likely develop what can be called informal knowledges, allowing them to engage in sustained study in an effective manner, or allowing them to intuit attributes of a problem and then choose among many prospective pathways forward in their research program. But appeals to logos and pathos are not the exclusive domain of expert-to-expert rhetorical activities. For experts who are not sanctioned, gaining credibility can be challenging.

Research on these challenges of gaining credibility as a non-sanctioned or professional expert allows us to further refine notions of expertise.[11] Indeed, this is imperative as to not reduce assessment of expertise or expert status to professions, systems of accreditation, or other social bodies that function to authorize knowledge-making and knowledge itself. But this example opens a common tension and that is the rhetorical work of establishing expertise and deploying technoscientific expert knowledge appropriately within the constraints of democratic forms. As we see, expert status here is distinct from one's expertise upon entering public, deliberative spaces. Thus, logos and pathos are implicated, alongside ethos.

Ashley, a citizen scientist, has a doctoral degree in cultural studies and works in a library. Ashley's insights nicely summarize many of the issues raised by citizen scientists concerning more expansive notions of expertise. Working in a library is particularly interesting for the question of assessing and understanding expertise.

Libraries, repositories for our collective human knowledge, are at the heart of how stored knowledge is organized, assessed, and understood. "I've worked in libraries throughout my academic career," Ashley tells us. Working in libraries, Ashley adds, "brought me into contact with a lot of different experts" citing not merely those in STEM, but those working in "media production, documentary film, audio stories, podcasts," and other creative fields. Encountering a range of experts in this capacity has allowed Ashley to cultivate an ability to help researchers across disciplines understand and manage their own projects. It was, however, not only libraries that afforded Ashley with the experience to build this kind of expert capacity in project design—it was Ashley's engagement in documentary filmmaking, which involved "learning about project design and being able to work with a bunch of different communities." "Being able to manage a bunch of different ideas, people, tasks, to make something as complex as a ninety-minute documentary film, has taught me a lot," Ashley explains, adding they are able to bring together "lots of different projects and help people as they design their own projects whether that's utilizing something like backwards design or being able to just sort of think through the different audiences that people might want to engage with, as well as just data workflows and things like that." Library experience—where "you bump up against lots of different people and [try] to sort of solve problems or create opportunities"—provides Ashley with a kind of expertise that is at once specialist, in the ways that those who cultivate our knowledge and projects have specialist capacities, but also generalist in the sense that this person is not so constrained by disciplinary conversations as to have the kind of trained incapacities (in Burke's 1984 Veblenian echo and Toulmin's 2001 stance against disciplines) that limit the possibilities for projects and, central to the present study, collaborations. Here the ways in which academics conventionally conceive of expertise, along disciplinary lines, is challenged by a familiar resource: librarians. Librarians' expertise is a crucial reminder that even the distinctions between specialists' and generalists' understanding of expertise are fraught. Interestingly, this problem seems to apply to a degree for the humanities, including fields such as rhetorical studies, where the broad application of specialist knowledge appears to follow a generalist model. It is in fact, as disciplinary journals themselves attest, not so. Yet the forms of specialist knowledge in these disciplines are configured in a sufficiently different manner from many technical expertises such that assessing the humanities and arts, and to some extent the social sciences, can be challenging.[12] Citizen science, Ashley notes, provides an opportunity to reevaluate notions of expertise: "I think citizen science is one of the more important avenues for this kind of work and also the sort of possibility for really exploring and reevaluating, and perhaps even sort of reaffirming a conception of expertise that's not based in credentials and not just based in publications and not just based in the sort of usual metrics of it." In addition to expanding conceptions of expertise beyond what are typically academic markers of credibility (e.g., credentials and publications), Ashley reminds us of expanding narrow formulations of expertise within academic settings.

In cases where sanctioned professional experts work with communities, there is also space for different forms of expertise and credibility. Ellis holds a bachelor's degree in environmental sciences and master's degree in marine sciences, the latter with considerable multidisciplinary training in not only the sciences but also economics and policy as well as communication studies. As a scientist, Ellis works on conservation, including socioeconomic concerns related to conservation. When asked how they became an expert, Ellis provides an especially interesting account, saying that they developed their expertise with a community-based program by "actually conducting that research myself. And so with my own ears, my own eyes, my own hands, if you will. That research project, in particular, that I was working on involved interviewing well over one hundred different people and asking them about how conservation affects their daily lives. . . . I think that my expertise, if you will, is drawn from not only researching and reading and writing a lot about community-based conservation practices, but actually talking to the people on the ground that do these practices and analyzing their answers." Here it is worth noting that the kinds of expertise relevant to a problem may be many, and sometimes it may rather be experience or even emotional responses that citizen scientists bring to research that are, although not technical expertise, still valuable. Ellis tells us that "community science is just as valid as [professional science]; well, not always, of course, but can be just as valid as science that is conducted by an expert as long as it's getting proper oversight." While, Ellis continues, "I'm over here preaching about how expertise takes a lot of time, a lot of intimate connection and familiarity and comfort with the topic. . . . I believe that that can be translated into community science programs really well." Citizen science in its varieties does not need to fit conventional conceptions of professional expertise, and, as among scientists, not everyone needs to have the same kind or degree of expertise insofar as it is restricted to content or skills-based knowledge. Other forms of knowledge offer crucial insights into research, even when not given by an expert. Ellis summarizes this matter nicely, explaining when we phoned for the interview, "I was actually doing an analysis of my community science data when you called. And I'm looking at this data and I'm thinking, 'This is really, really interesting information that I would not have access to if it weren't for my volunteers.'" Ellis continues, "Even though I'm the 'expert,' and my volunteers aren't experts, I'm able to provide them with that practical expertise to collect this data and then give it to me." Ellis's point reaffirms the importance of different forms of knowledge in the practicing of expertise. Further, Ellis notes the importance of expertise in a community of scientists and citizen scientists, each bringing different knowledge and skills. "I can use my data expertise to interpret it and provide it to the people that need it to regulate our protected areas," explaining that this is significant because "I think that continuum, that connection of knowledge between the scientific experts, the community scientists and the regulation makers, I think is really, really important and that expertise looks different at every stage of that process."

Expertise, Ethos, and the *Polis*

A common refrain about the rise of experts sees the encroachment of technoscientific rationality onto civic deliberative discourses and spaces. Indeed, citizens may bear the brunt of such disasters, including financial burdens. How can technical experts weigh in appropriately to deliver their specialized knowledge and assessment of situations while acknowledging the necessary space for public deliberation in a functioning democracy?[13] Consider River's experience and the importance of local knowledge in scientific research. River is a birder, studying bird crossings near roads and highways and also seasonal sightings. Having worked in forestry for thirty-five years, River was able to study land, ecosystem, and other forms of forest management. "I've had extensive weather training because of fires and stuff like that, so I can see how a lot of those different things relate to what's out there, like whether it'd be clear-cut logging or oil and gas development," River explains, further noting how this is an integrated form of knowledge: "That's what I can see: how it relates or how it can relate." Although River has been involved in citizen science in one form or another since the 1980s, in response to the question "Would you describe yourself as an expert in your work as a citizen scientist?" they say "No." River explains that they are, however, an expert in different venues, saying they have been an expert witness, not because of degrees or other credentials, but because River was advanced as an expert with local knowledge, quoting the rationale for their expert status: "The local people knew more about what's going on than the scientists." River, although seemingly reluctant to adopt the "expert" label, explains they in fact have knowledge that is exceptional. River explains that by being in the location and birding for many years, "you begin to understand much more about what's local than somebody who comes cruising in to do three studies and then leave.""I don't consider myself an expert," River says, "but some other people might." Gaining this expertise—my attribution, not River's—involved several years of what we might describe as minimal engagement in birding, but about a decade ago River, following their retirement, became highly engaged. River explains that part of their development was volunteering, taking photographs of and helping to categorize birds; and seemingly as their expertise developed, so too did the social acknowledgment of said expertise. Soon River found themselves giving presentations to introduce birding to new citizen scientists and helping with bird identification among professionals. When training new citizen scientists, River reminds them of why their contributions are essential: "I explain to people how important it is that the scientist can't ... do it all themselves. Contributions are very important. It might help scientists with things like population trends of birds or you're concerned about climate change, just things like that. You might observe something that scientists can't possibly observe because they're not there." Here, then, an interesting feature of the expertise citizen scientists might bring is by virtue of their local knowledge and their experience in that location. Fundamentally this presence reminds us of the critical importance of early work by naturalists. The generation of scientific knowledge relies on not only observation but also the recording, the inscription of phenomena. Citizen scientists, in this respect, are most certainly participating in research and cultivating forms of expertise related to observational field work and the written practices of science.

The perceived threat of narrowly limiting such local research to professional spheres is the overtaking of democratic decision-making by technocratic decision-making. Specialization does have a function in expert knowledge, but its limits are real and expert knowledge is more expansive than specialization. Knowles (2011) shows that certain forms of specialized individual expert knowledge are critical when integrated into a team's knowledge. Knowles makes the case by examining the complexities of policy-making practices, looking at the dire consequences of not heeding technical or scientific expertise in development of policy and risk management. Indeed, following Beck's (1992) caution about the complex nature of risk, the importance of multidisciplinary teams of experts in addressing risk becomes critical, particularly because risk is inequitably distributed.[14]

Expertise and the Role of Phronesis

For Aristotle, we noted above, phronesis is among the intellectual virtues and is the ability to make well-reasoned decisions.[15] Central to the case Majdik and Keith develop, phronesis is not defined by its outcome but rather by the "disposition" of a person to make good judgments based on a set of both universal and particular criteria for prudential reasoning.[16] Here they follow closely the Aristotelian tradition of virtue ethics. Majdik and Keith (2011a) make a compelling argument for the rhetorical function expertise serves in sociopolitical spheres, or the "dimension" of expertise that functions in democratic or civic spheres of discourse. In their formulation, expertise is "a kind of authority," which has serious implications for engagements with a democratic public (371–72). As a form of authority, expertise "stands in contrast to liberal democratic values; at its core, a democratic polity depends on its ability to keep a check on authority" (371–72). Here Majdik and Keith articulate authority and expertise as coterminous, and thus ask what boundaries should be applied to the authority of expertise when it intersects with sociopolitical decision-making. Their aim is to understand expertise a form of argument or argumentative practice, and they do so by arguing that expertise cannot alone be described merely by "possession of knowledge" (374). Further, Majdik and Keith claim "expertise does not reside in the knowledge or experience of the arguer"; rather, it is the skill of responding appropriately (373). Expertise in this sense may be read as answering Engeström's (2018) call for distributed notions of expertise rather than strictly individual models, as here expertise is constituted by the situation and not solely an individual actor. In Majdik and Keith's own words, "Phronesis emerges less from the actual enactment of prudence[17] as it does from one's disposition and ability to act prudently should a situation arise that demands action," which means that in a reconfiguration of the concept, "expertise can be said to be a function of an argumentative tekhne that originates from a phronetic sensibility relative to specific problems expertise addresses" (376, 377).[18]

Indeed, practice is well-grounded rhetorically in Schön's articulation of reflective practice in professional-client encounters. Reflection occurs with respect to not only the situation but also the audience. According to Schön (1983, 295; quoted here at length to underscore the rhetorically aware conception of the reflective practitioner):

> Just as reflective practice takes the form of a reflective conversation with the situation, so the reflective practitioner's relation with his client takes the form of a literally reflective conversation. Here the professional recognizes that his technical expertise is embedded in a context of meanings. He attributes to his clients, as well as to himself, a capacity to mean, know, and plan. He recognizes that his actions may have different meanings for his client than he intends them to have, and he gives himself the task of discovering what these are. He recognizes an obligation to make his own understandings accessible to his client, which means that he needs often to reflect anew on what he knows.

A reflective practitioner thus is set against the authoritarian expert, with presumptive claims to black-boxed knowledge, strategic distance, and an expectation of deference (300). Instead, a reflective practitioner is someone who has some "recognition that one's expertise is a way of looking at something which was once constructed and may be reconstructed" (296). Expertise here thus seemingly operates within the parameters of reflection-in-action among practitioners, which is partially informed by notions of tacit knowledge[19] and is set against models of technical rationality.

In such a reconfiguration the audience, too, must shift its stance to one that no longer demands certainty, and instead cultivate modes of inquiry that, although they may not be able to assess expertise, can assess the expert as willing to engage in deliberation. In doing so, however, both practitioner and audience engage in a different relationship. Practitioner and audience do so knowing they "bring to their encounter a body of understandings which they can only very partially communicate to one another and much of which they cannot describe to themselves," and thus must work together in inquiry (296). Fundamentally this offers a reconfiguration of the contract between experts and nonexperts (or, following Schön's terminology, "clients") to a "reflective contract" (302). Schön's central preoccupation is the relationship between research and practice. His configuration of the reflective practitioner reconciles the two through reflection-in-action, not merely as implementation or practice of research findings; rather, reflection-in-action offers a transformative practice, modifying the situation (309). But he continues, explaining that research (or, perhaps, we might say knowledge or engagement with ways of knowing) does not necessarily occur within this context of practice in ways that are relevant to the practitioner's reflection-in-action. Cultivating capacities, he explains, to better engage in reflection-in-action constitutes a kind of "reflective research" (309). What he then describes includes

1. "frame analysis," wherein "frames determine [practitioners'] strategies of attention and thereby set the directions in which they will try to change the situation, the values which will shape their practice," and reflection on these frames is essential to understanding what tacit knowledge or frames are at work in their thinking (309);
2. "repertoire-building research," which allows for "familiar situations, cases, or precedents" to be accumulated (315);
3. "research on fundamental methods of inquiry and overarching theories," which describes research that examines the methods and theories "that some practitioners have learned to use as spring-boards for making sense of new situations which seem, at first glance, not to fit them" (317–18);
4. "research on the process of reflection in action," which requires experimental research studying the cognitive, cognitive-emotional, and social or situational circumstances for reflection in action (in problem-solving contexts) with attention to fear of failure and other social conditioning (320–23); and
5. understanding that "researchers and practitioners" are in a collaborative process together, requiring itself a reflexive stance of reflection in action. (Schön 323–25)

Some of the general principals laid out by Schön and their relationship to features of phronesis have not gone unnoticed (see Kinsella 2010).

An entomologist with expertise in evolutionary biology, Riley provides a good case study in the complex reflective activities, including integration among researchers and practitioners, necessary to become expert. Riley explains that becoming expert has two parts: the first, "reading a hell of a lot, taking courses," and, the second part, "experience." In this account, experience provides a space for how knowledge can be constructed and reconstructed, following Schön. Here the common distinction between acquiring knowledge and experience is well defined. Riley explains why these two parts of expertise are required: "The reading and the courses don't tell you what it's like on the ground. You got to get out there and do it. And find out what fails and what works and learn for yourself about that." Indeed, this is an interesting presaging of what will later be discussed in terms of knowledge building for students. For now, we might simply acknowledge that what Riley describes is not merely acquiring knowledge but identifying "what works," a matter of application of knowledge. At the time Riley was trained, they did not have much opportunity for multidisciplinary education. Since then, a considerable amount of what we can call multidisciplinary work has helped to shape Riley's research. The multidisciplinary scope of this research is a function of the problem space; Riley explains that for their line of research one must work with people who experience on-the-ground problems. Such problems entail material and economic conditions, which require different forms of expertise to solve problems.

Riley's work, however, also includes working with researchers across the humanities and social sciences, and such work has proven challenging as it requires further reflection on questions of expertise and assessment of expertise. Riley argues that respect and epistemic humility are critical features of expert knowing, saying that "it takes time to develop that respect often for people who have different ways of knowing. I think that's something that I have learned a lot about that kind of being able to judge expertise when you have a certain way of knowing. Like an equation will let you know something. . . . An anthropologist says, 'My way of knowing is to sit—try for two years and watch what happens.' Then the sociologist says, 'My way of knowing is to do a survey. We're to do focus groups.' Those are different ways of knowing."

Understanding disciplines with quite different epistemologies and methods, and assessing expertise across those disciplines, proves challenging. Framed here as a matter of respect, invoking notions of goodwill toward other expert audiences, cross-disciplinary research is at once important and challenging. However, the difficulties in training students to work across such epistemological divides is all the more challenging, conceptually and practically. Riley trains students in a domain of biological sciences, but this field relies on an apprentice model that involves being in the lab. Taking time to leave the lab, stepping outside of one's discipline, even with incentives, can be challenging as it comes "at the expense of getting the work done."[20]

Expertises, Rhetoric, and Doing

Further complicating how we might understand aspects of expertise that are not explicitly knowledge or practice based in common renderings, Rice (2015) offers an account of what she calls "para-expertise," also drawing from Polanyi's (1966) concept of tacit knowledge, defined as *the experiential, embodied, and tacit knowledge that does not translate into the vocabulary or skills of disciplinary expertise*" (119, emphasis in the original). Collins and Evans (2007) might categorize this kind of expertise under "ubiquitous" expertise, which is, recall, the kind of expertise people gain from growing up in some society. Rice examines how college students interpret certain architectural features of university campuses, explaining how their embodied and affective experiences allow them to develop tacit knowledge that is important to experts planning campuses (Rice contrasts her idea of para-expertise with Collins and Evans's contributory and interactional expertise). Rhetorically such reframing allows for better accounting of how expertise is done, where the experience of students is key to better expert practices among those designing campuses; Rice explains that such a perspective "shift[s] attention to how para-expertise can lead to problem-posing" and allows for such para-experts to "articulate exigencies by validating real needs, problems, and experiences" (Rice 2015, 131; cf. the experiential epistemology of Hartelius 2011, 123; see also, on digital place and phronesis, Xiong-Gum 2018). Problem-posing can be compared with Schön's (1983) notion of "problem-setting" versus "problem-solving," the latter a preoccupation of practitioner-experts in the tradition characterized as technical rationality, and the former as an alternative in Schön's epistemology of reflection-in-action, which "accounts for artistry in situations of uniqueness and uncertainty" (165). For Schön, problem-setting is part of the sense-making process that professionals must engage in real-world problems, and he explains it thus: "When we set the problem, we select what we will treat as 'things' of the situation, we set the boundaries of our attention to it, and we impose upon it a coherence which allows us to say what is wrong and in what directions the situation needs to be changed. Problem setting is a process in which, interactively, we *name* the things to which we will attend and *frame* the context in which we will attend to them" (40, emphasis in the original).

As alternatives to problem-solving, both Schön's problem-setting for professionals and Rice's problem-posing for para-experts offer alternatives to a solely technical rationality. Importantly, such an articulation allows for a different configuration of engaged nonprofessional experts than cited in the cases of Wynne's sheep farmers or in my own work on citizen scientists (Kelly 2016; Kelly and Maddalena 2015, 2016; Mehlenbacher and Miller 2018), where in the former case, specialist knowledge was held in the form of local specialist knowledge and in the latter the formerly nonexpert in measuring radiation contamination gain specialist skills and knowledge to become specialist experts. Para-experts, instead, may create "strategic expertise alliances" (132), Rice explains, with interactional or contributory experts. Although these features—experiential, embodied,[21] tacit—might suggest an individualist model of expertise, Rice argues that "expertise is less an individual quality than it is a description of the activity of posing problems" as the construction of the problem space and rhetorical situation will influence who might respond and how (122).

DeVasto (2016a) similarly frames expertise as an activity, refiguring the Collins and Evans normative model of expertise through Mol's (2002) multiple ontologies to understand expertise through such an ontological framework as *doing*. Key to understanding DeVasto's conception of expertise as doing is her understanding of what an ontological, rather than an epistemological, stance toward expertise offers, especially an ontological stance that accounts for the multiplicity of experience and expertise. DeVasto draws again from Mol to explain that the explanatory punch of multiple ontologies is not merely one of multiple perspectives but rather of multiple realities. In such a formulation, DeVasto explains, analysis can focus on the enactment of expertise rather than the kind of knowledge we might normally ascribe to expertise. Through this framework, DeVasto offers an "expertise of doing" (378). To illustrate this expertise of doing, she examines the case of the 2009 earthquake in L'Aquila, Italy. The earthquake had been predicted and debated among scientific and other experts, but ultimately was dismissed until the tragic event took more than three hundred lives. Complicating these events were poor communications by experts about the threat of seismic disturbance, which led citizens to believe they did not need to take precautionary measures. Normally, when such tectonic activity seemed to increase, individuals in the community would take to sleeping in cars or outdoors, as the buildings in the community were known to be risky in the event of an earthquake. Because citizens had been reassured there was no threat of a significant event, they did not follow standard protocol. Later, following an investigation and legal trail which found six scientists guilty of misconduct—a sentence that would later be overturned—questions arose about who was responsible for such poor communication of risk. Refiguring the Collins and Evans model of expertises, DeVasto shifts them to distinct ontologies—the "interactional ontology," the "contributory ontology," and the "referred ontology"—ultimately arguing that the L'Aquila case lacked sufficient interactional representation, and thus suggests the importance of multiple interacting ontologies in such public scientific deliberation and policymaking.[22] Majdik and Keith (2011b) also offer an explicit articulation of expertise as practice-centric. Recall, however, their work (2011a) on the role of phronesis in expertise, where they do not measure expert practice through results, but rather through the ability to identify possible means of expert engagement. The important work Majdik and Keith do for a rhetorical concept of expertise is allow for the tension between global and local heuristics for just who counts as an expert. Importantly, they underscore that the outcomes of expert decision-making are not as useful a measure in many cases as the decision-making itself, which may be shaped by not only expert practice but often broader social conditions (e.g., resistance to masks during the COVID-19 pandemic).

To achieve a practice-centric perspective, Majdik and Keith set aside a strict focus on the epistemic aspects of expertise—not discounting knowledge, however—to focus on how distinct situations might invoke global heuristics. The perspective also considers how the practice of expertise can resolve those for a particular problem or situation through expert involvement, although this is not the measure of expertise. In such an account, they argue, "we may approach 'expertise' as a practice that relates to forms of life," qualifying that "practice-oriented concepts of expertise would add to, rather than replace, the epistemic versions of expertise" in an effort to define expertise broad enough that it is not restricted to a subset of specialists with particular knowledge (2011b, 278). Drawing on Wittgenstein's (2009) language games, practice ("obeying a rule"), and concept of "family resemblance," Majdik and Keith (2011b) explain that practice involves individuals acting within social and material conditions. "Enactment of expertise involves understanding of that problem and its solutions that could, as possibility rather than necessary actuality, be projected on to future cases," they write, then explaining this means that expertise conceptualized as "practice" "is the process of being able to articulate reasons for—thus operating under a requirement to socially validate and legitimize—an individual's enactment of expertise" (279). Thus, moral dimensions, too, are implicated.

Expertise and Moral Responsibility

Moral responsibility is foregrounded in Pietrucci and Ceccarelli's (2019) study of the L'Aqulia earthquake. Examining communications that took place by the six scientists—who, recall, were tried, convicted, and later acquitted—only days before the disaster took place, Pietrucci and Ceccarelli reject the grounds for the acquittal on a rhetorical-moral, not legal, basis. They argue that while imprisoning scientists is not what they would advocate, the scientists do bear responsibility for the failure to appropriately communicate risk to the citizens of L'Aqulia. The rationale they offer for this is that these scientists, and indeed any scientist who participates in the public sphere,[23] are those who possess knowledge to correct misunderstandings. Critically, such experts have an ethical obligation to do so. "To earn the trust we invest in them," Pietrucci and Ceccarelli write, "scientists must draw upon a full rhetorical ethos grounded in moral values (arête), goodwill (eunoia), and practical judgment (phronesis)" (98). Indeed, they argue that scientists must understand their role beyond a restricted technical sphere because, despite efforts to maintain boundaries between discourses in technical and public spheres, scientists' "boundary work does not absolve them from their ethical responsibilities as members of a polis" (101).[24] Notably, in this account, Pietrucci and Ceccarelli reject DeVatso's (2016a) model of interactional ontologies as response to communication failures such as those in L'Aqulia and place responsibility on scientists (those scientists who Collins and Evans might call the "contributory experts" or in the cases of expertizing that DeVasto might examine in a "contributory ontology").

In their analysis, Pietrucci and Ceccarelli point to three critical junctures where scientists *should have*, they argue, intervened to correct misleading or factually incorrect information. Crucially, their argument posits the role of scientists as having a duty to enact responsible rhetorical citizenship. Among these duties for scientists is to participate fully in the public sphere concerning their areas of expertise. Thus, they invite "scientists and experts to think of their role in society as going beyond the mere deployment of their expertise in the technical sphere" because scientists' expertise, "when confined to the technical sphere, can be vulnerable to distortion, exploitation, neglect, or misunderstanding in the public sphere" (122). Miller's (2003) investigation of how risk analysis, which demonstrates that in some scientific disciplines there is a propensity to substitute expertise in place of ethos, anticipates the concerns raised in Pietrucci and Ceccarelli's account of expertise as constrained to technical discourse. In the development of risk analysis as a field, expert opinion became important to manage indeterminate data dealing with the probability of events. However, expert *opinion* was not entirely persuasive to those engineers. Miller examines the 1975 "Reactor Safety Study" (sometimes called the "Rasmussen report" after its director) for the Nuclear Regulatory Commission. In her analysis, she explains that the Rasmussen team understood the challenges involved in deploying the more subjective and qualitative data of expert opinion and relying on ethos over the favored logos. Rhetorically, this can be accomplished through an interesting move, Miller (writes:

"What might in other situations be central to an ethical appeal—affiliation, prior success, masterful expertise—in science and technology must be treated as logos, as factual evidence, attributes of the technical situation rather than of an advocate in a rhetorical situation" (184). Not to be fooled by this masquerade, Miller reminds us that this is ethos in the form of a denial of ethos (185). In this way, appeals to authority might be made from a presumptive position of expert status, although the experts assessing these strategies were skeptical.

For the public, however, Miller argues the Rasmussen report "combines a traditionally impartial or objective scientific ethos with a paternalistic authority" demonstrated in document's executive summary, which marks the "narrowing of ethos to expertise" (193, 194). In her account, the transformation of ethos in expert-to-expert discourse is to logos, and in expert-to-public discourse, the transformation of ethos is restricted to emphasize expert knowledge, through which, she remarks, "ethos becomes merely a shadow of logos" (195). Miller's major contribution is in her reformulation of the relationship between ethos and expertise. For Aristotle, she reminds us, ethos is deployed in those situations where expertise is not among the available means of persuasion, such as when communicating to a nonexpert audience. For Miller, "expertise stands in for ethos":

> An ethos of expertise—that is, an ethos grounded not in moral values or goodwill, or even in practical judgment, but rather in a narrow technical knowledge—addresses its audience only in terms of what it knows or does not know. The diminution of *arete* and *eunoia* in an ethos of expertise has a specifically rhetorical effect, because these qualities are relational in a way that expertise is not; similarly, the transformation of *phronesis* to *episteme* diminishes the practical, or relational, dimensions of knowledge. Without *arete* and *eunoia*, there is no basis for agreement on values or for belief in the good intentions of a rhetorical agent. (201)

Miller elaborates by explaining the consequences that such an ethos on trust a nonexpert audience may have. By exploring the distinction between ethos and expertise in a technical setting, with both an expert and nonexpert audience, Miller provides important distinctions between the concepts of ethos and expertise. Notably, she underscores the moral component of ethos through arête and eunoia, and through practical judgment as phronesis. An ethos of expertise, however, that is grounded in "narrow technical knowledge," as Miller describes, is, in addition to an impoverished conception of ethos, an impoverished conception of expertise. Like ethos, expertise is a habituated activity, one grounded in convention, and one that demands certain ethical commitments. If ethos describes relational aspects—those that characterize attributions of expert status and not expertise per se—then an ethos of expertise too requires arête, eunoia, and phronesis. Indeed, in Miller's case of the Rasmussen report, the lack of ethos or its transformation into a kind of logos-lite, was not especially persuasive among experts or publics. Miller's articulation of why an ethos of expertise fails to be persuasive identifies the crucial role of one's moral comportment. Experts, broadly, are implicated here and the rationale for their engagement as participants in both technical and public spheres applies equally well. The role of experts in relating complex risk information to publics is a well-treated concept in rhetorical studies of science. Indeed, the question of how publics can assess experts—determining if the expert-rhetor has goodwill toward the audience (eunoia) and virtue (arete), along with practical wisdom (phronesis)—is one that merits continued attention. In Miller and in Pietrucci and Ceccarelli, another critical concern for rhetoricians is the relationship between what is often called the technical sphere and the public sphere with respect to moral duties. Because experts have knowledge that others do not, and because said knowledge—be it technical in nature or, even, the knowledge generated in the humanistic traditions—has implications for how everyday people live, experts cannot be absolved of their moral duties of intervention, as Pietrucci and Ceccarelli describe.

Expertise and Experts

The rhetorical account of expertise developed here conceives of expertise as problem-responsive, where expertise is a matter of knowing-why in a particular situation. Experts must know what means are available, what specialist knowledge is relevant, and how to deliberate and act judiciously. Ultimately, the rhetorical perspective offered in this chapter establishes expertise as comprising knowing-that (episteme and experience), knowing-how (skill or techne), and knowing-why (phronesis). But the addition of phronesis inflects expertise with an important quality, and that is a moral one. Here I depart somewhat from the rhetorical focus of DeVasto's and Majdik's take in that I wish to concentrate on aspects of the individual performance of expertise in addition to the situational configuration demanding expertise to resolve some urgency. Enactment of expertise, how we decide to combine our prior knowledge or expertise or skills to a particular problem, is habituated in us through socially sanctioned and valued engagements.

By suggesting something can be habituated in us, our mind is implicated. While we certainly are constituted and constrained by the world we inhabit, which seemingly all schools of thought on expertise acknowledge to some degree or another. Those worlds comprise linguistic, social, cultural, and indeed material aspects, often in complex configurations. However, for many fields studying what it means to be an expert, measures of expertise rely not only on engagements within those complex configurations but also the individual capacities to act in a manner we might call expertly.

In the next chapter, psychological perspectives on expertise and experts will help chart the commonest understandings of expertise. While I have attempted to fairly account for each field's central contentions and contributions, my accounting is in favor of those matters seemingly most related to rhetorical theorizing. From this multidisciplinary exploration of expertise, we learn that expertise and experts are not so easily identified and grasped. Rhetorically this is fascinating because the very terms of expertise and expert status are constantly negotiated and renegotiated across these fields. The search for precision in definition, criteria for assessment, and social understanding of the role of expertise are all contingent on argument. Understanding expertise as a rhetorical phenomenon, then, is more than an effort to demarcate boundaries between experts, expert knowledge, expertise, and so on, and nonexperts or publics. Rather, understanding expertise as a rhetorical phenomenon is understanding the varying and changing social terms of expert status, the formulations of expertise, the configuration of expert knowledges and skills, and the value of experts in our deliberative spheres. Also required is understanding of the capacities that allow for experience to be structured in such a way that it can be honed in an individual. Key insights into this process come from psychological studies of expertise. Presaging some of these insights of rhetorical traditions, memory plays a vital role in individuals' training—both in terms of their technical proficiency as rhetoricians but also in their practical moral character—a position that we can trace through Cicero and Quintilian, but becomes most powerful during the medieval period, notably in the work undertaken in monastic traditions. Such memory is not merely how much we can recall but a more complex sense of how we cultivate forms of knowledge through our individual recollective practices. As early as the *Dissoi Logoi* we find treatment of this idea of the artificial memory, and the distinction between memory for words and memory for things, memory places, and memory as a key tool in invention unfolds throughout antiquity to the medieval period.

3

Expertise, Psychology, and Memory

Socialization is a rhetorical process that we have charted in the preceding chapter but, as yet, the psychological principles articulated there have not been integrated with rhetorical thinking. In the study of expertise, to be sure, there are both social and individual aspects. For the latter, rhetorical theory has tools to understand this process of becoming expert, notably in the tradition of rhetorical memory, and does so attendant to the social body. Memory, the following account demonstrates, is a powerful concept for understanding experts not only in terms of practice-based scenarios with well-defined parameters, but also in those situations that require generative or inventive work to solve complex problems. Although the strategies of memory can indeed be characterized in psychological terms, explaining some of the mechanisms that humans rely on to store information, studies of memory also implicate how inventive abilities are tied to the rhetorical worlds we inhabit and, further, how those abilities are grounded within ethical norms. Although memory is a concept that recurs in the study of expertise, numerous modern conceptions of the term often focus on mechanical processes of memory as retention rather than an artistic, inventive process. It is the latter sensibility of memory, as an inventive process, where rhetorical traditions might slightly shift our focus to how ethical judgment is implicated in expert thinking. All told, memory may be among the most important rhetorical principles in the development of a rhetorical theory of expertise not merely for its conceptual relationship to how knowledge, practice, and experience are integrated into expert thinking, but, also, critically how these concepts are shaped within the social and individual ethical comportment of an individual.

Numerous fields of study,[1] including philosophy, psychology and brain sciences, education, and sociology, theorize the factors shaping cultivation of expertise, such as domain-specific knowledge, socialization and discourse communities, social environment, individual and social interplay, temporality, material environment, brain plasticity, memory, mental representations, experience, talent, forms of practice, instruction and training, adaptability, and intentionality.[2] Missing from many accounts of expertise is a sustained investigation of the rhetorical dimensions of expertise.[3] Rhetorical thinking, however, is woven into many fields studying expertise. In the introduction to *The Science of Expertise*, Hambrick, Campitelli, and Macnamara (2017, 2) write that "the Ancient Greeks laid the foundation for the contemporary debate over the origins of expertise." Citing Plato's *Republic*, where Socrates argues for differences in individual aptitudes,[4] and Aristotle's rebuttal and postulation that it is in fact experience that accounts for knowledge, these authors identify debates that have also shaped the Greco-Roman rhetorical tradition. Unabated since antiquity, "the pendulum has swung between the view that experts are 'born' and the view that they are 'made,'" Hambrick, Campitelli, and Macnamara write (3). For the psychological sciences, the debate now centers on degrees of influence among genetics as "born with"[5] and acquired traits and environmental influence as "made from."[6] Although psychologists offer insights into the processes of memory that merit some consideration, it is rather the more social and rhetorical constitution of memory that is of interest here. Moving from the individual to the social body, rhetorical memory helps characterize not only how memory functions in the operation of what we can call expertising, but, also and critically, our understanding of how epochs shape what we understand to be expertise itself. It is with this Blumenbergian lens that attending to the developments in rhetorical memory, especially our metaphors for the operations of memory, might tell us much about the ethical norms and values entailed when we speak of memory.

Consider, by way of example, how Carruthers (1990) illustrates the function of memory in expertise when she examines accounts of two exceptional individuals from two distinct periods: Albert Einstein and St. Thomas Aquinas. Comparing accounts of these two individual's minds, Carruthers illustrates that modern preoccupations are concerned with imagination, originality, innovation, and so on. Prior to this, memory was understood somewhat differently, but still functioned as a central feature of great minds capable of invention. Several commentators remarked on Einstein and St. Thomas's memory. Einstein's collaborator at Princeton in the 1930s, Leopold Infeld,[7] reports on Einstein's creative genius and singular attention to his work. Bernardo Gui, who wrote a hagiography not long after St. Thomas's death, and Thomas of Celano, who Carruthers explains was a witness at St. Thomas's canonization hearing, both comment on St. Thomas's impressive compositional feats owing to his powerful memory. Passages for both men underscore their "qualities of genius," although phrasing marks different preoccupations with their respective contemporary thinking, Carruthers notes. Each case illustrates the qualities that are often attributed to both genius and notable expertise including "profound singlemindedness,[8] a remarkable solitude and aloofness," "intricacy and brilliance of the reasoning, and its prolific character, its originality" (3). Carruthers argument is that, although the process descriptions vary, the processes of thought, the "compositional activity" or "the activity of thought," is indeed similar for both Einstein and St. Thomas (4). In antiquity, Galen, too, persuaded his audience partially on his mastery of prior works, along with a healthy dose of appealing to his experience as a medical practitioner (König 2017). Memory is an inventive process predicated on more than a "good" working memory and rote memorization (reports of Einstein's "poor" memory referring to rote memorization), but, critically, organizing and recollection. But such practices are highly individual in nature, in terms of how such organizing and recollection occurs, while also being constructed within intellectual predecessors and communities, which is to say they are socially informed and rhetorical in their nature. For medieval thinkers, including those commentaries on St. Thomas, inventory and invention are crucially interrelated, as one's contribution must demonstrate both an awareness and "inclusion of major elements from his predecessors" (Carruthers and Ziolkowski 2002, 22). In her accounting of the qualities that mark these two extraordinary individuals as surpassing their peers are echoes of the qualities we use to describe experts, including single-mindedness, deliberate practice, and considerable knowledge; however, we also, critically, find those qualities of reasoning characteristic of phronesis, which makes them social and rhetorical characteristics.

Ultimately, a rhetorical account of expertise does not rely solely on a knowledge-based (knowing-that) model or practice-based expertise (knowing-how) model. Rather, a rhetorical account of expertise critically conceives of expertise as problem- and situation-responsive; or, in other words, knowing why a given response is most appropriate.[9] Both common models of expertise, the knowing-that and knowing-how, implicate each other, each providing some weighting and attention to the other. Claiming that expertise requires knowledge and skill seems rather uncontroversial, if a little reductive. A model of expertise where knowing-why is the central focus is likely to be uncontroversial, too. In such a model, expertise still requires knowing-that and, often, knowing-how, but adds attention to the capacity to survey one's knowledge, understand a particular situation, and apply some heuristics to deliberate and adjudicate[10] on the best enactment of expertise. Such a capacity is central to the enactment of expertise.[11] What the present description lacks is a nuanced accounting of how we know why—how does one survey a set of universals to respond to particulars, what kinds of inventive strategies are deployed in finding a solution, how do experts deliberate on options, and so on—and what is meant by *why*.

Expertise Through Practice, Deliberate and Otherwise

Chief among those fields interested in the concept of practice is psychology,[12] and decades of study offer fascinating insights into how "practice" is theorized and operationalized. Beginning with a practice-based definition, Gobet (2016, 5) offers a definition of an expert as "somebody who obtains results that are vastly superior to those obtained by the majority of the population."[13] Here experts are understood in terms of their performance, and there is some granularity in the idea of experts, as some may surpass even other experts ("super-experts").

Among the most often cited research in the psychological sciences on practice-based expertise is Ericsson, Krampe, and Tesch-Römer's (1993) work on "deliberate practice."[14] Ericsson, in his long career studying expertise, distills much of this important work on deliberate practice in a popular book, co-authored with science writer Robert Pool, titled *Peak*. Summarizing the research conducted by Ericsson and colleagues, practice is a difficult concept and often misunderstood, they argue, citing the example of a driver who has been practicing driving for twenty years versus a newer driver who has been practicing for only five years (Ericsson and Pool 2016, 13). Such perceptions of practice as repetition of activities advancing our capacities is wrong, they argue, arguing that "research has shown that, generally speaking, once a person reaches that level of 'acceptable' performance and automaticity, the additional years of 'practice' don't lead to improvement" (13)—further, sometimes individuals' capacities worsen as their intentional efforts to improve stagnate. To understand why this happens, Ericsson and Pool explain there are different forms of practice: naïve practice, purposeful practice, and, without equal, deliberate practice. In their model, naïve practice is a kind of unreflective repetition, whereas purposeful practice involves having precise goals, being focused, receiving feedback, and crucially, pushing oneself outside one's "comfort zone" (14–17). However, "trying hard isn't enough"; rather, it is another form of practice, deliberate practice, that requires participation of already-experts to assist in training and development (25).

Deliberate practice, Ericsson and Pool[15] argue, requires something more than purposeful effort. Instead, deliberate practice "requires that a field [be] reasonably well developed" such that establishing individual top performers from average or even "good" is possible, and they cite music, chess, and gymnastics as examples (98). Further, "deliberate practice," they explain, "requires a teacher who can provide practice activities designed to help a student improve his or her performance" (98). Similar to *purposeful* practice, deliberate practice involves having defined goals, being focused, receiving feedback, and moving outside one's comfort zone, but it also requires working in a specialty where others have developed "effective training techniques," building skills over time with the assistance of an expert, and producing "effective mental representations" (99). Aristotle made a similar point about habituating in the cultivation of virtues. Sherman (1989, 179) explains that, in *Nicomachean Ethics* (2.1), Aristotle provides an example of learning to play the lyre "by practicing not merely with persistence, but with an eye toward how the expert plays and with attention to how our performance measures against that model," and further that "without the instructions and monitoring of a reliable teacher, a student can just as easily become a bad lyre player as a good one."

An expert in computational neuroscience, Casey notes how specialized this area of research can be, and that such specialized expertise was acquired through doctoral and postdoctoral work. Similar to other experts, Casey explains that continued work in the field was an important part of developing expertise, particularly "in a multidisciplinary team, [which] has led me to sort of take greater ownership of that particular piece of the pie and develop even more expertise." Once again, we hear a familiar refrain: "So it's a combination of training and then on-the-job experience." On-the-job training involved working in a large pharmaceutical company with a research and development group. Such research is "inherently cross-disciplinary"; Casey explains that their teams include a variety of different kinds of scientists, clinicians, and others. Casey offered a particularly insightful explanation of becoming expert, beginning with personal experience and moving to training students: "My own experience in developing my expertise had a lot to do with reading on my own, discussing critically the findings with more kind of senior colleagues, and then actually trying to rigorously confirm or disconfirm these ideas empirically." First having students read, then discuss what they have read with their supervisor, develop a hypothesis, and try to confirm or disconfirm findings is the model on which Casey relies. In the field, it is "usually more often disconfirmation or a lack of a conclusive outcome than it is confirmation," Casey notes, "so that helps contribute to this better calibration of uncertainty." Again, we learn that reading, gaining the knowledge as episteme, is important, but so too is the acculturation process, discussing findings and learning the norms of the discipline, which is enacted and modeled through apprentice-ship and perhaps deliberate practice.

While deliberate practice offers a compelling narrative about the importance of environment, dedication, and practice itself, some researchers in the psycho-logical sciences are skeptical of such totalizing claims. Other models attend to the ways in which genetics, or "innate abilities,"[16] and phenotypical interactions may afford or constrain one's potential to attain expert status, along with the environmental and individual practice-based factors previously detailed.[17] Although deliberate practice seems to play a role in how a given individual is able to cultivate their skills and abilities, when comparing multiple individuals, delib-erate practice seemingly loses some of its explanatory punch. Something else seems to be going on that allows some individuals to learn at a faster rate or achieve a higher level of skill. Their answers for what is missing in the deliberate practice model are complex, but likely include aspects of the individual as well as their situation. In sum, the nature-nurture debate is over, and the discussion has moved to how different variables interact and the degree to which they have influence (for one such model, which integrates "wisdom-based experience," "creative expertise," "practical expertise," and "analytical expertise," see Sternberg 2017, 421). Although some competing claims exist within the field, it appears that psychological studies have a reasonably satisfying conception of expertise, although practitioners in other fields may disagree. As Gobet (2016) argues, there is no single mechanism or singular approach, and studies across various levels of mechanisms are crucial to understanding such a complex phenomenon. Psycho-logical studies provide experimental, data-driven approaches to studying cognitive mechanisms that are indeed difficult to articulate.

Dylan, a scientist who works with citizen scientists, earned degrees in fisheries and marine science and is currently completing their doctoral studies. Much of their research is related to costal and marine habitat and species. Dylan explains that their path toward expertise was not simply set in motion by way of their studies, but that interest and motivation were key. Before university studies, Dylan tells us, "the interest I had already in the natural world and science about the natural world" was important, adding that "a lot of it was just self-taught with books, I watched TV shows, made sure I was involved in different small groups of people going out, so then you start to learn." Later, at college, fieldwork became an important part of Dylan's training, and this translated into the kind of work they would need to perform in citizen science projects: "It just helped with knowing kind of what to look for in a bit more detail. And if there was something unusual going on . . . stuff that you wouldn't normally think about would matter, but when you know it matters then." Working on citizen science projects, too, has afforded Dylan with ongoing learning opportunities, as the projects one is involved with ensure one is "always learning more." Dylan provides an example by way of a story. They began working with a graduate student, whom we will call Sandra, who wanted to launch a citizen science project. The project involved testing fish caught by fishers to examine them for traces of plastic. Sandra enlisted Dylan to help test procedures and Dylan soon discovered they did not perhaps have the necessary proficiency in plastic identification. In that case, Dylan explains, they did not perhaps have a "good level" of expertise, but in other areas, where they have more experience, such as habitat surveys, Dylan explains that their "expertise increases." Here we already see that practice may involve certain individual features, such as motivation, but also relies on a social configuration for learning how to be expert.

Expertise and Practice: A Psychological and Educational Account

One feature common among practice-based definitions, such as in psychological sciences, is to view experts as superior cases, the exemplars in a field or specialty. In these accounts, the long histories of humans who excel at some task provide a rationale to begin the study of experts. Chess players have been the subject of some of the most compelling research on experts.[18] Their specialty is highly constrained and their results measurable and, importantly, comparable with others. What are called Elo ratings allow for the systematic ranking of players and thus identification of experts among them. Athletes with exceptional physical abilities and psychological fortitude are common among examples or groups studied to reveal the mechanisms of expertise. As with chess players, there are often ranking systems that allow individuals to be assessed relative to their competitors' capabilities. Classic musicians, too, allow for such systematic study and assessment of relative competency. These are the experts that provided the participant pool for much of the foundational research in the psychological sciences on expertise. Experts, however, are also found in those areas where the criteria for establishing the legitimacy of status is somewhat more difficult.

When we begin our journey of becoming expert, we must ask questions about how much time we spend practicing; who we practice with, and their pedagogical strategies, commitments, and responsive feedback; our community of other experts or expert learners; our levels of experience; our skill decay; our material and social environments; and indeed the composition of our bodies and, within those bodies, the plasticity of our brains, affect, and memory. For example, as an expert in knowledge representation, Noor explains that studying at a university where research in this area was well-researched was foundational to their becoming expert. "And that's what everyone was talking about," Noor says. "There were discussion groups, there were projects going on. So that's where I sort of soaked it all up. It was an amazing experience." Noor explains that at the time of completing their undergraduate studies, there was more flexibility in programs, and they were thus able to complete courses in the humanities, sciences, and mathematics. Later, multidisciplinary work was part of Noor's research program, but they encountered difficult-to-overcome challenges including academic systems of evaluation. Different granting mechanisms are valued differently across the disciplines. Various forms of instruction, too, are valued differently. Noor's discipline is focused on content delivery, but learning requires a richer framework, and Noor argues that "communicating well is part of doing science. You have to be able to collaborate. Collaborate with your students. Collaborate with your colleagues." To accomplish this, even Noor's classroom instruction values "independence" and "creativity." Noor may be an interesting example of an expert learner in the face of a system or pedagogy that does not, in fact, support expert learning to a large degree.

To understand why it may be that conventional educational models do not necessarily create experts, a model of how one becomes an expert is useful. Dreyfus and Dreyfus (1986), in an effort to demarcate the boundaries between human cognition and machine learning, outline stages that indicate integral moments in becoming expert. This model has been widely discussed and is useful here as a kind of heuristic for understanding the process of becoming expert rather than as a particular model to which we must commit. In the model, it follows that people do not simply acquire some form of theoretical or rule-governed knowledge (knowing-that) and suddenly become capable of applying that knowledge (knowing-how). Instead, Dreyfus and Dreyfus explain, normally a person moves through stages on their journey toward becoming an expert. There are at least five of these stages of their model, which can be roughly demarcated as novice, advanced beginner, competent, proficient, and expert. These stages are marked by several notable transitional characteristics, including the abilities to manage both context-free and situational knowledge, becoming more experienced, and moving from analytical to intuitive decision-making, the latter of which is reserved for experts (50). "An expert's skill," they write, "has become so much a part of him that he need be no more aware of it than he is of his own body," and this means that in the normal course of their duties *experts don't solve problems and don't make decisions; they do what normally works*" (30–31, emphasis in the original). Although experts sometimes must deliberate, they acknowledge, experts come to act with a "fluid performance" (32; cf. Csikszentmihalyi 1990, 1996, 1997) even in those more complex situations that are dependent on intuition (as arational, not irrational)[19] because conscious rationale thought in fact impedes or regresses expert performance (36). Gobet (2016, 4) suggests these definitions are at odds with one another, "almost the opposite definition of expertise"; however, if we understand expertise as requiring practical wisdom, we will see that one is able to both engage in something like fluid performance while also addressing problems at one's upper limits. Experience must be paired with that effortful movement at one's upper limits, otherwise there is a danger that experience might carry the pretense of expertise without any of the functional benefits. It is not especially difficult to see in the case of doctors how this might occur in the routine socialization of their practice, as they move from novices to experts—with years of ordinary medical problems and procedures, they may lose their edge to address new cases.

Medical experts of different varieties offer compelling examples for how one might become and continue to be expert. Consider Hansaa, a biological anthropologist who has expertise in public health emergencies, including pandemics. After completing doctoral studies in a program designed to be cross-disciplinary, Hansaa gained further experience in teams that span not only academia but also thinktanks, industry, and the public service sector. Like other experts we interviewed, a combination of training in graduate studies and then continuing to work on research programs provided the foundational training and practice to establish expertise. But Hansaa, with considerable on-the-ground experience through the application of research to community problems, offers an important insight to how we think about expertise and experience. "I think it is important, if the project team has been put together, to kind of just sit down and talk about what you've done," Hansaa explains, adding that practical questions to assess expertise and experience might include "What have you done, literally, not just in terms of the main subject areas, but what have projects looked at? Have they been more scientific? Have they been in a certain part of the world? Have they been within certain socioeconomic groups?" These questions provide "a sense of what you've all worked on and where you might be able to kind of leverage somebody's . . . not necessarily expertise, but perhaps kind of experience of something that you need to bring into your project." The distinction between experience and expert here is notable. To become expert, Hansaa argues that one must be "on top of the literature all the time to know if it's changed." Understanding the literature requires understanding both the research with which one agrees but also an ability to assess literature that one does not agree with while maintaining a fairness in assessment. Hansaa explains that one must be apprised of the current literature and be open to engaging those you do not necessarily agree with in a serious way to advance your own understanding. "You can't just dismiss anybody whose work contradicts your own view of something," Hansaa argues. When training future experts this is a crucial point for Hansaa, because "you shouldn't be trying to turn out students who are duplicates of yourself." Practically this means that you may train students, future experts, who subscribe to different schools of thought, and even contrary, Hansaa adds, to one's own, and then the mission is rather not of mere content sharing but helping one learn. Training should involve having students, for instance in the sciences, ensure that "their research methodologies are sound. They're doing their research in the right way." Such an approach, to Hansaa's mind, is crucial, even if students fail: "It may be then that they do the study they want to do and find that the results contradict what their initial view was," and this also causes reflection on the prior research informing study design, which is an important capacity for gaining expert understanding of a research area.

Psychological Models and Social Critiques

Expertise that occurs in seemingly well-constrained environments, such as a chess match or an athletic tournament, or even driving a car, provide a well-structured context to study expertise. However, increasingly in the twenty-first century, when we call on experts we are asking them to address highly complex problems.[20] Although a large number of experts are highly specialized in subfields or sub-subfields of research, many must work across subfields or in teams comprising individuals from entirely different domains. These kinds of complex, multidisciplinary situations provide an interesting environment to consider expertise. Experts in these teams must develop capacities to work across domains. Outside of highly specialized research contexts, this kind of configuration of multiple experts is common. Another challenge in such contexts may be that the team must quite rapidly learn how to assess the parameters of a problem and which experts might respond to particular aspects of the problem. One could consider, for example, a medical team deployed in response to an epidemiological crisis, where one must work with not only various kinds of medical experts (across subdomains) but also other actors in crisis situations, such as those scaling up infrastructures for a field hospital to raise capacity for patients. Expertise, too, may be found within an effected community, which has important knowledge about which professional scientists and public health experts do not have insight. Or, in another case, perhaps medical staff would need to quickly coordinate and work with emergency responders in the event of a natural disaster. For example, consider a direct hit by a tornado on a hospital and the need to evacuate patients; indeed, this was the situation faced at Mercy Hospital Joplin (then St. John's Regional Medical Center) in Joplin, Missouri, after a deadly tornado in 2011 caused extensive damage. We will see, then, that to understand the individual account alone is not explanatory of expertise.

At the individual level, dedication, grit, sheer volume of hours—famously giving us the ten thousand–hour rule[21]—and dedication to one's craft are all implicated in the psychological research on expertise. From a rhetorical vantage, however, it is rather the social conditions that seem especially critical to one's development of expertise, particularly in addressing intractable problems. Among relational models of expertise, the social configurations that shape our beliefs and knowledge about the world are central to our understanding. In such accounts, expertise and expert status are not merely the *telos* of significant knowledge and practice, but an ongoing negotiation between the orator who claims special knowledge, skills, or proficiency and audiences who might have occasion to judge such claims. In the study of science, the social dimensions of experts and expertise have been well covered, notably by figures such as Harry Collins and Robert Evans, Shelia Jasanoff, and Yrjö Engeström. Although each account offers different insights into the nature of expertise, each draws our attention to rather distinct features of the practice-based literature. Assessing one's capabilities is regulated by conventions of a particular specialty. For example, medical professionals require a high degree of socialized practice to fully become integrated and understand the field. Although specific knowledge and skills certainly underlie much of the work in medical professions, additional norms and values operationalized through discourse strategies, genres of communication, expectations for professional conduct, affiliations, reputation, and so forth illustrate some of the more social functions that also help establish the relative credibility of one's claims to expertise or expert status. For researchers interested in these social dimensions of expertise, the stakes of their research are not merely taking the epistemological giant of science down a notch, but are also crucial to understanding why experts and their expert knowledge sometimes fail.[22] What much of this research has shown is that science is a situated practice constrained by a number of social factors that shape what constitutes science at a particular time.

Notably, psychological models have been challenged by those working in the Sociology of Scientific Knowledge's "Studies of Expertise and Experience" (SEE). Chief among those developing models of socialization have been Collins and Evans (2007) and their model of expertise described above. Collins (2018) distills the position of SEE on expertise, defining an expert as "someone who has been socialised into an expert community or specialist group," further clarifying the distinction from psychological models of expertise saying that "SEE makes no distinction between acquiring an expertise and becoming socialised into a group" (352). Socialization is an important departure from traditional models in Collins's view, where "most definitions of expert are tied up to the notion that an expert is more right or more efficacious than a non-expert" (352). For Collins the crux of the matter here is that a SEE model of expertise allows for disagreement between experts. He tells us that "two persons who have been socialised into a specialist group and are therefore experts in that specialty (say, GW [gravitational wave] physics) can disagree fiercely with each other without analysts having to conclude that one of them must be less of an expert than the other" (352). Importantly for our purposes here, Collins adds a parenthetical note stating, "One might expect such accusation"—that one party is less of an expert—"to be part of the rhetoric of arguments among experts themselves" (352). This is a crucial point because Collins recalls the change of scientific thinking and related expertise over time, citing the example of alchemy or astrology and the subsequent fields of study that depart quite radically in thinking on related subjects. In Collins's view, these experts of now defunct concentrations can be examined "without there being any question of the experts in the discredited group no longer being counted by the analyst as having been experts in that specialism" (352).

Collins's articulation of sociological models is not entirely at odds with aspects of psychological understandings. Ericsson and Pool's (2016) account of expertise, for instance, does not simply assert a model of expertise centered on individual agency. Rather, the focus on the individual as a unit of study instead reflects disciplinary norms, and indeed a further investigation into how those disciplines acknowledge such limitations reveal their more comprehensive articulations of expertise. For example, in deliberate practice, the role of a teacher or what, echoing Lave and Wenger (1991), we might call a community of practice, is essential to the very conception of *deliberate practice*, as set apart from *purposeful practice*. In such models, the role of a broader system is not dispensed with, but rather the current site of investigation focuses on individual mechanisms. Indeed, the concept of deliberate practice offered by Ericsson and Pool relies on socialization as a mechanism of training.

Haru offers insight into the complexity of becoming expert in both psychological and social terms. Explaining that, like many technical undergraduate degrees, a degree in math often involves non-math electives, including humanities courses, Haru also notes that their fascination with the arts was nourished during their coursework. During graduate studies, Haru chose to continue pursuing courses outside their own discipline. "I've always loved art, and patterns, and design, and tilings," Haru says, explaining that although they had many years away from those studies while pursuing degrees in mathematics, eventually they were able to bring these two areas together in doctoral studies. At an early age, before their teen years, Haru would stare at paintings. Since then, their doctoral studies and subsequent career have been filled with the intersections of math and art. Because of the relation of math and art in Haru's work, they are always attentive to new ways the two intersect. Following a strategy supported by the psychological literature, Haru tells us, "Lucky people aren't actually more lucky. They're just always sort of more open-minded and more receptive to coincidences and connections. . . . I try to live my life that way." On becoming expert in a technical sense, as well as a social idea of becoming expert, Haru explains (echoing others) that it is a process of immersion in the research and the research community: "By seeing talks, and looking at papers, and reviewing, and occasionally writing papers, you just kind of—you get immersed in that community and generally become aware of what's out there." Rhetorical inquiry aligns quite well here in discussions of the socialization into becoming expert, particularly when the socio-cognitive models of genre learning (see Berkenkotter and Huckin 1993, 1995; Ding 2008) are compared with Collins and Evans's articulation of socialization within a specialty. Novices, recall, can gain competency in some specialty by either naïve or purposeful practice on their own. For example, one might begin learning to play the guitar by picking one up and following a training manual to learn some of the basics. Although this strategy may work for a time, it will not allow them to excel in the same way that one might were they to follow a deliberate practice model. According to this more demanding routine, the student would work with a mentor or attend a music school, which perhaps implicates the kinds of tacit knowledge we obtain through immersion within a community of practice. That is to say, overall gains in music and athletics attributed to deliberate practice are dependent on the socialization within specialty and the growth of knowledge in training practices in those specialties over time.

Mental Representations and Memory in Expertise

It is through the Ciceronian Latin transformation and form of practical wisdom, *prudentia*,[23] that we might understand more about the means by which one might cultivate this virtue and also, perhaps surprisingly, expertise. Cicero's *De inventione* offers four cardinal virtues, which includes prudence. Prudence is defined as "the knowledge of what is good, what is bad and what is neither good nor bad"; Cicero then explains how prudence is constituted, noting that "its parts are memory, intelligence, and foresight. Memory is the faculty by which the mind recalls what has happened. Intelligence is the faculty by which it ascertains what is. Foresight is the faculty by which it is seen that something is going to occur before it occurs" (*Inv. rhet.* 2.53.160, quoted in Carruthers 1990, 65; see also Yates 1966, 20).[24] St. Thomas Aquinas adapts notions of prudence from Cicero and also from the *Rhetorica ad Herennium* (see Cicero, pseudo 1964), emphasizing the ethical character of *prudential* reasoning, and still maintains a closely related conception to Aristotle's phronesis (Zagzebski 1996, 212).[25] Aquinas extends a critical discussion of the mechanisms for cultivating prudence by making meaning of experience over time: that is, memory. Citing Cicero's account of *prudentia* in *De inventione*, Carruthers (1990, 66) argues that prudence's temporal nature includes "memory being of what is past; intelligence of what is; foresight of what is to come." From here, explaining Aquinas's understanding of prudence, Carruthers cites him at length, which I have reproduced here with her explanatory asides and abridging (on Aquinas's conception of *prudential* vis-à-vis Aristotle's phronesis, see also Zagzebski 1996, 212–14):

Tully [*Inv. rhet.* 2.53] places memory among the parts of prudence. . . . Prudence regards contingent matters of action, as stated above [*ST* II-II 47.5]. Now in such matters a man can be directed . . . by those [things] which occur in the majority of cases. . . . But we need experience to discover what is true in the majority of cases: wherefore the Philosopher says [*EN* 2.1] that "intellectual virtue is engendered and fostered by experience and time." Now experience is the regular of many memories as stated in *Metaphysics*, [1.1], and therefore prudence requires the memory of many things. Hence memory is part of prudence. (*ST* II-II 49.1, quoted in Carruthers 1990, 66–67)

Crucially here, Carruthers explains, Aquinas understands experience resulting not from "many memories of the same thing" but rather (pace Aristotle) *ex pluribus memoriis* or "from several different memories" and, thus, "the memorial experience that founds prudence is not iterative but concatenative" (67). For Aristotle, experience precedes multiple forms of knowing, including episteme and techne (see *Metaphysics* and *Posterior Analytics*) and phronesis (see *Nicomachean Ethics*). Such experience requires, further to accumulation, an ongoing process wherein "certain connections [can] be broken and reassessed in the light of anomalies" (Sherman 1991, 192), and experience requires time, as Aristotle explains we do not remember the present but only the past, and thus tells us that the mechanism or organ by which we experience time is also the mechanism by which we are able to remember (*Mem.* 1.449b24–30, in Sorabji 1972). Continuing Aristotelian thinking on the habitual character of recollection, and building on Augustine's understanding of the "trinitarian nature of the human soul (memory, intellect, and will)," Aquinas advances memory as *habitus*. In doing so, he "makes it the key linking term between knowledge and action, conceiving of good and doing it. Memory is an essential treasure house for both the intellect and virtuous action" (Carruthers 1990, 64). Memory provides a helpful link between the kind of socially recognized and rationalized forms of such wisdom discussed by rhetoricians of science studying expertise and the capacities of individual experts that seems to preoccupy psychologists. The idea of memory here merits some further attention, namely for its distinctions from how we understand it today and how medieval thinkers conceptualized of memory.

Memoria is a process by which the five senses' perceptions are transformed into impressions called *phantasma*.[26] As the impressions are divided to be stored, they are placed in locations created by the schematics such as the treasure house or memory palace, or, in other configurations, the Guidonian hand or the seraphic angel (see Carruthers and Ziolkowski 2002, especially pages 5–7). Key to understanding the approach here is that these locations are rendered as images visible at a glance in the mind's eye. Such images might take the form of scenes, but never more than what one might look at in one picture, not a series. Images, further, had to be distinct, and this might be accomplished by rendering them in an emotionally charged manner or in a situationist vein, out of context. Carruthers and Ziolkowski (2002, 13) explain that these requirements are "deliberately grotesque and fanciful, in a manner we now consider characteristic of medieval aesthetic." Once the impressions have been divided and stored in these image-based schemata, the memories are formed and can later be retrieved. The systems of organization allow for retrieval from different locations in memory. Retrieval serves the broader function of recollection. Building the treasure house, a process of *diairesis* (Latin *divisio*) in reading, is critical to the enterprise. It is, however, the process of recollection that is essential to the questions of expertise, and especially to practical wisdom in expertise. Carruthers's (1990, 69) explanation of how memory underlies practical wisdom is insightful: "The ability of the memory to re-collect and re-present past perceptions is the foundation of all moral training and excellence of judgement." On the question of recollection and expertise, we can see the importance of experience informing different forms of knowledge that allow for a kind of flow state. That is, the ease that seemingly marks much expert performance is afforded by the well-built, stocked, and organized treasure house. Having a particular individual organizational system for the purposes of retrieval in invention is an important aspect of expertizing.

Recollection works as an inventive, compositional activity and is not limited to the ability to locate and repeat information in one's treasure house. Rather, at least during the medieval period and central to the accounting of memory, recollection relied on one's ability to integrate and compose from memory. Drawing from texts, medieval memory composition relies on their authority in establishing an author's own, and this required that authors carefully incorporate those texts into their own fabric; this was done partially by the reading and then, later, by recollecting texts. In the recollection we find again a connection to the moral value of this process, as Carruthers explains, situating this process of reading and recollecting in the ultimate compositional activity of amplifying. In this sense, the *res*, the sign, is the meaning to be found within a text (not, as we might now discuss, in authorial intention or in the reader). This meaning "which must be amplified and 'broken out' from its words, as they are processed in one's memory and re-presented in recollection" (191). Crucially to the questions of expertise and moral knowledge, Carruthers's explanation of this process illustrates how memory is implicated in one's moral character: "Amplifying is an emotional, image-making activity . . . and it is just this quality that makes it ethically profitable. More importantly than growth in knowledge, reading produces growth in character, through provisioning—in *memoria*—the virtue of prudence" (191). It is not simply that having a "good memory"—so commonly conceived of as high retention and sometimes used as an ableist conceit—makes one compose in these ethically profitable ways. Rather, memory is an inventional act, and the process of recollection is not merely rote. Memory is, further to this, constructed through our engagements with others intellectually and rhetorically. In this way, we can reject simplistic notions of "good" memory to understand the important ways that individuals engage in inventive work and how distinct processes and experiences open new horizons for broader ethical imaginations, and epistemic horizons, too.

Through this inventional account, Carruthers's recovery of the medieval craft of memory is essential for a rhetorical formulation[27] of expert thought.[28] In her model, memory is not relegated to a "mere" technical canon, paired with delivery and contrasted with invention, arrangement, and style as more generative processes or "philosophical" canons (13). As we learn from Carruthers, *memoria* of the medieval period offers a complex site of study that will illuminate cognitive, rhetorical, ethical, and social dimensions. Each of these dimensions, further to our inquiry on expertise, has much insight to offer. First, the very concept itself must be given some rough description. Memory, following Aristotle,[29] was divided into two acts: the first, *memoria* or *mnesis*, which is the process of storing information (through the process of *divisio*, wherein information is broken into sections and then arranged in some order that will make it easy to recover or recollect), and the second, *reminiscentia* or *anamnesis*, the process of recollection, which is central to the inventive activities of composition (Carruthers 1990; see also Carruthers and Ziolkowski 2002). Although more common usage of memory conflates the two, keeping these delineated explains the way ancient and medieval thinkers conceived of memory. In an illuminating metaphor from the *Epistulae Morales ad Lucilium*, Seneca likens the compositional activities in which we engage to the production of honey: "We ought to imitate bees, as they say, which fly about and gather [from] flowers suitable for making honey, and then arrange and sort into their cells whatever nectars they have collected" (Carruthers 1990, 192; translations her own from *Ep.* 84, Reynolds ed.). Memory is a dynamic and inventive process. Carruthers and Ziolkowski (2002, 2) explain that for medieval thinkers, "memory-making was regarded as active; it was even a craft with techniques and tools, all designed to make an ethical, useful product," including sermons, hymns, meditations, and the like.

Modern Memory Craft

To recover memory as something more than a mechanical process or parlor trick, Carruthers (1990, 19) sets out to describe the deeply cognitive and inventional conception of memory in the medieval period: "The proof of a good memory lies not in the simple retention even of large amounts of material; rather, it is the ability to move it about instantly, directly, and securely that is admired." The ability to move about largely relied on the visual aspects of memory,[30] and the metaphors used to talk about memory reveal its complexity with respect to knowing. Here we should not understand aspects of memory to be rigid schemes per se. Indeed, Carruthers reminds us of the complexity of these memory systems. She explains that "all ancient mnemonic advice" account for the idea "that any learned technique must be adapted to individual preferences and quirks" and requires training of the mind (64). Indeed, noting the prescriptions of the early handbook tradition, Carruthers (2006, 210) explains that "like memory itself, specific mnemonic techniques are ultimately idiosyncratic and arbitrary; this was consistently recognized in the common admonition never to use others' schemes except as a guide, in order—through practice—to make one's own." Whatever the cognitive constraints of our memory systems, it seems, from centuries of compositional activities, that whatever those affinities of our wetware, individually crafted memory schemes offer a powerful ability to constitute and reconstitute knowledge through inventional practices. That is to say that, although the techniques of memory almost certainly have underlying cognitive structures that have reasonably ordered schema,[31] these structures operate with enough finesse that they do not cultivate universal moralist principals. Interestingly, much of what we learn from ancient and medieval memory techniques bears a relationship to what we now know about memory. In antiquity and the medieval period,[32] the limits of short-term memory were understood; and indeed, the ideas of building, for example, memory palaces or encoding information in numerical schemes helps move information into long-term memory.

Today such trained or "artificial" memory is unlikely to be given as a topic, but we can see that compositional practices rely on certain forms of memory training.

Here the practice of *memoria ad res* provides a framework to consider the inventional capacities of artificial memory. In the interview data offered in the following chapters, experts talk about the inventional and compositional practices that allow them to not only practice their expertise but also become expert, including by reading articles and attending conferences. Such engagement with the field allows those becoming expert to build further structures as well as acquire new knowledge. Gobet (2016) notes that although the mnemonics born of ancient rhetorical training of memory are less frequently used today, they still provide tools for improving memory, and, further, they tell us much about how memory might work in expertise.[33]

Common among research on expertise is attention to de Groot's (1965) foundational analysis of chess players in the 1940s. De Groot's work illustrated that expert chess masters were capable of quickly remembering and then reproducing configurations of chess boards, which was quite different from nonexpert capabilities. Further, chess masters seemed to show a capacity to restrict their representations to the most likely configurations of a board. In describing how this process works, de Groot also calls attention to the ways that mental images are used in the process of memory:

> The rapid insight of the chessmaster into the possibilities of a newly shown position, his immediate "seeing" of structural and dynamic essentials, of possible combinatorial gimmicks, and so forth, are only understandable, indeed, if we realize that *as a result of his experience he quite literally "sees" the position in a totally different (and much more adequate) way than a weaker player.* The vast difference between the two in efficacy, particularly in the amount of time to find out what the core problem is ("what's really cooking") and to discover highly specific, adequate means of thought and board action, need not and must not be primarily ascribed to large differences in "natural" power for (means) abstraction. *The difference is mainly due to perception.* (306, emphasis mine)

In further elaboration, de Groot's account of memory recalls the ancient idea of memory as a visual process: "It is no accident that the word 'seeing,' as used here, stands both for perception and (means) abstraction. The two processes tend to fuse together; they are difficult to distinguish" (306). The ancient refrain continues to echo in de Groot when he writes, "It is above all *the treasury of ready 'experience'* which puts the master that much ahead of the others. His extremely extensive, widely branched and highly organized system of knowledge and experience enables him, first, to recognize immediately a chess position as one belonging to an unwritten category (type) with corresponding board means to be applied, and second, to 'see' immediately and in a highly adequate way its specific, individual features against the background of the type (category)" (306, emphasis in the original; see other editions or translations, which may use words such as "store"). De Groot further adds, "The gist of the preceding discussion might be summarized by saying that a master *is* a master primarily by virtue of what he has been able to build up by experience; and this is: (a) a schooled and highly specific *way of perceiving*, and (b) a system of reproductively available methods, *in memory*" (308, emphasis in the original). Although de Groot's work was foundational to the study of memory and expertise, Chase and Simon provided a more detailed account of the mechanisms that seem to allow for superior recall among experts.

Chase and Simon (1973) developed the theory of "chunking."[34] Their chunking theory "makes three key assumptions: (a) chunks are a single storage unit of both meaning and perception, and are retrievable from LTM [long-term memory] in a single act of recognition; (b) STM [short-term memory] is limited to about seven items . . . , for both experts and non-experts; and (c) players of all skill levels learn new information at a relatively slow rate" (Gobet 2016, 31). Gobet explains the significance of this work, writing that Chase and Simon performed a similar experiment to de Groot to test recall, likewise finding better recall in more expert plays. Chase and Simon also designed their study to include a copy task, which provided an explanatory function for how experts' perceptions and memory performed better than novices' (30). Ericsson would later discuss this in purposeful and deliberate practice. The strategy used by chess masters is one wherein they "chunk" information, aiding in their recall. Crucially, chunking information suggests a strategy where some knowledge base makes certain configurations of information meaningful, making it more readily recalled. From a rhetorical perspective, it is not only the cognitive mechanisms that underlie memory that are interesting, but also the way memory allows for recollection to respond appropriately to a given situation. Practically, this means that knowledge is implicated in expertise, but, further to this point, knowledge is not wholly constitutive of expertise. Expertise requires a capacity for action, deliberation, judgment, and decision-making, and it is in that action we find knowledge enacted to respond to a particular problem or situation. In terms of the cognitive operations, the current research on memory would likely not astonish medieval and ancient thinkers. It would not be surprising, that is, in terms of how material should be chunked or that there are important limitations of memory (or that limitations on memory are themselves important) which memory practice can work around to some degree. Indeed, Carruthers and Ziolkowski (2002, 4) note that when cutting up material into *divisiones*, the "segment should be 'short' (*brevis*), no larger than what your mental eye can encompass in a single glance or *conspectus*" (see also Carruthers 1990, 80–107).

Indeed, memory schemes in antiquity and the medieval period are not, Carruthers cautions, those computational models prevalent today. Instead, memories were understood to be formed by sensory input as well as affective dimensions at the individual level.[35] That is, pathos[36] is an essential component in the formulation of memory, and thus also in the habituating or training of memory crucial to the cultivation of the virtue or prudence. This is why, too, event-based or experiential memory may be important, and also why spatial features such as memory palaces, topoi, and the like may be as well. Memories, too, are temporally bound, especially in Aristotle, who understood memory as a "re-presentation" as much as a representation, a kind of recollection of a previous experience situated in the present; thus, "recollection was understood to be a re-enactment of experience, which involves cogitation and judgement, imagination, and emotion" (Carruthers 1990, 60). "Experience—memories generalized and judged," Carruthers writes of Aristotle's account of ethical excellence in *Nicomachean Ethics*, "gives rise to all knowledge, art, science, and ethical judgement, for ethical judgement, since it is based upon habit and training and applies derived principles to particular situations, is an art, and part of the 'practical intellect'" (68–69). Aristotle understands emotions to be essential to the proper and full reasoning required to understand and remember information in a way we might recollect it rather than merely reciting it.

Memory provides some insight into the individual capacities required to become expert. In the sense of memory developed here, the techniques used to arrange, store, and retrieve experiences are what the ancients would have spoken of as artificial memory. As a kind of trained memory, we find similar techniques in studies of experts, which is unsurprising. Indeed, as Yates (1966, 369–73) reminds us, memory as a technique did not simply vanish but was reformulated and influenced the development of scientific thinking by figures as prominent as Bacon in his *Advancement of Learning*. Unsurprisingly, too, the role of excited emotions in memory recounted by ancients is present today, too, in how we understand the capacities of memory. Memory, however, provides more than insight into how individuals can retain and retrieve experiences that allow them to surpass themselves, to become expert (see: Bereiter and Scardamalia, 1993). Rather, memory, as a part of prudence, provides a crucial link between individuals and the communities within which they are embedded. As a part of prudence, memory is not merely what is past, but the kind of integrated and arranged experience (including of exceptions) that might help us learn, with the benefit of reflective intellectual capacities, how to discern and then how to act. In this discernment and action, the choices are, for an expert, not merely technical but, relationally as their status demands, also contingent, rhetorical, and moral in their composition. It is, then, not that a good memory makes us good, but that how we choose to integrate experiences with respect to the communities within which we are embedded helps us deliberate and make good decisions.

Setsuna works in educational technologies after having studied psychology. When asked how they believe that they became an expert in their area, Setsuna provided an answer that echoes what we learned of experience as constituting memory: "I think there are two ways. The first way is I have to structure my knowledge. What I mean is I have to organize my experience because if you don't do that your knowledge is fragmented and I think that's not good." The refrain of memory craft is again called on, noting the importance of experience not simply as experience, but as it becomes incorporated within existing structures of understanding. The second way Setsuna believes one becomes expert is "you have to make your network . . . and you have to make network with some people in your area and also outside your area." Building a network is an important feature, Setsuna tells us, because one must both "absorb new ideas" from the network of others as well as engage in the individual work, which is a matter of "time and practice." Setsuna explains the process of becoming expert to their mind: "First, you have to focus on one area you're really interested in," which means for training that "the first thing I want my students to do is to find their interests and to focus." The work "to clarify your interests" is central, Setsuna tells us, because often when interests are initially described the "goal is not specific." Having a goal to help direct interests is important because that allows one to begin to establish parameters for success. Once interests have been focused, "You have to take time and effort to practice," Setsuna explains. "And most of the time, the practice is not comfortable." Interestingly, this idea that practice should push people, move them outside of their comfort zone, recurs in discussions of expertise. It seems quite a sensible position, too, that as one practices "you have to observe what you feel, and you have to adjust to situations, and you have to seek advice from people." Setsuna provides an account that brings together many of the ideas explored in this book, from the deliberate practice model from psychology, to the importance of memory craft, as well as the role of community in the process of becoming expert. Along with Setsuna, many of our participants illuminate these ideas, and the next two chapters consider responses from self-identified experts, offering a wealth of insights across fields.

4

Professional Researchers on Expertise

At the outset of this project, I was interested in how researchers in multi-disciplinary collaborations evaluated the expertise of prospective collaborators. Asking experts questions about how they understand the idea of expertise, broadly, and how they understand themselves becoming expert framed later questions about collaboration. What has been especially interesting about the findings is that, although knowledge and experience are often cited as central attributes of expertise, experts also shared insightful comments about the importance of other qualities in those experts with whom they have collaborated. In the Likert-based list of adjectives[1] that survey participants were asked to complete, most participants strongly associated or associated "knowledgeable," "experienced," "informed," "skilled," and "credible" with experts. Other terms we might associate with credible through rhetorical traditions, including "ethical," "reliable," or "truthful," however, were more somewhat more moderately associated with experts. Although these associations are not meant to be statistically significant or wholly generalizable, they do suggest some noteworthy perspectives on expertise. It is reasonable to say that expertise is often viewed in conventional terms of knowledge and skills, correlating with knowing-that and knowing-how, given how survey participants scored these terms and what participants told us in interviews. But interview participants revealed much richer attributes that they look for when seeking out collaborators, including those qualities we might say demonstrate phronesis or knowing-why.

Ultimately, this leaves us with tough questions about the nature of expertise. Expertise and experts' rhetorical strength has been long principled on their exclusive access to knowledge or their ability to excel at some task or techne. The knowledge an expert has, in this model, is what provides them with special status, along with the ability to put into practice this knowledge. Knowledge and experience, together, are essential to the expert. "Essential," however, does not suggest that these two qualities are exclusive. Indeed, few survey respondents said experts did not need to be credible, or reliable, or have good sense. However, these terms do not appear to be consistently tied to an ethical sense. Indeed, roughly a quarter of participants said they do not associate "ethical" or "fair" with experts and a little over a third said they do not associate "good morals" with experts. Although there is, indeed, a compelling case for understanding why experts ought to attend to practical wisdom—complicated and messy as it is—there is a sense that the moral underpinning is not primarily the concern of the expert. Yet, these are important qualities. Indeed, experts themselves telling us they would rather work with individuals who demonstrate pro-social qualities, qualities requiring, among them, practical wisdom.

An important distinction here is that this study is especially interested in self-identified experts who have worked in interdisciplinary collaborations, as part of multidisciplinary or cross-disciplinary teams, or on trans-disciplinary problems. Experts with these kinds of experience are interesting because, to a degree, it would be exceedingly difficult to work in those spaces and not confront matters of expertise in a routine and perhaps more explicit manner than in, for example, a specialized lab. Offering some experience in navigating across the disciplines to understand who an expert is, what kind of expertise they have, and even how to understand one's own expertise, then, are important commonalities among our participants that offer particularly insightful thoughts on experts and expertise. What we are provided, then, are explanations from experts on their own heuristics for assessing expert status.

We will learn in the course of this chapter that there are various strategies to assess expertise, including looking at credentials or someone's track record of publications, grants, or projects to specialized skills. Another strategy is looking to someone's capacities to explain complex concepts to social markers of credibility, trust, and an ethical sensibility. First, the chapter examines definitions of expertise. These definitions were gathered from surveys and interviews, and they are inclusive of a range of different disciplines and career stages and paths. After formulating a rough definition of expertise based on responses, the question of how expertise is understood across disciplines or specializations is raised, with interesting implications for those in the humanities and social science working with STEM colleagues. In the following section, how one obtains expertise, as articulated by interview participants, is explored. In the concluding section, the question of how to assess expertise is examined through a combination of survey and interview responses.

Defining Expertise

Of the more than ninety participants who responded to our survey, when asked, "What are three words you associate with someone who is an expert or has expertise?" the commonest free-association responses were "knowledgeable" (appearing thirty-two times) and "experience/d" (appearing twenty-five times), and other commonly associated words included (appearing between four and nine times) "depth," "competent," "understanding," "insight," "published," "education," "intelligent," "skills," "confidence," and "rigorous." When asked, by way of an open-ended form, if there are other words one might associate with an expert or someone who has expertise, participants provided somewhat richer responses. A doctoral candidate in chemistry provides further insights about who is an expert: "I think of someone who has expertise as having an expansive knowledge base that encompasses their subdiscipline in great depth. Experts tend to have densely connected schema (cognitive structures that hold bits of information in long-term memory) that allow them to automate lower-order cognitive demands, and instead these experts have a wealth of problem-solving avenues to explore." Knowledge is understood as having "depth" in a "subdiscipline" and, importantly, a mechanism and function to which it might be applied. The idea of having "densely connected schema" invokes notions of memory or mental models, essentially some manner in which this knowledge is stored, and then some indication that there is a tacit mechanism for retrieval ("automate"). Both the mechanisms for creating memories and retrieving them function in service of allowing the expert to focus not on the knowledge, as episteme we might say, but instead on the "problem-solving," a function of practical wisdom (phronesis). "As a result," the participant continues, "I think true experts are those who are creative in their schematic connections, are generative in that they contribute to their subdiscipline, and are collaborative in pairing with other experts from disparate fields." Inventional aspects of this model of expertise are apparent, with the idea of "creative" connections and "generative" thinking in a specialization. In this account, these ends are principled in a very similar way to how memory was theorized, how research demonstrates a cultivation of expert capacities, and thus provides an interesting account of who is an expert and what it means to have expertise. A lead scientist in a machine learning project identifies expertise as awareness of one's limitations: "People who have expertise are willing to state when something is beyond their expertise," a refrain also recalled by an architect, who wrote, "[An expert is] aware that they may not know everything." Here is a familiar idea of knowledge affording the awareness of limits of one's knowledge, which provides yet another important feature of expert knowing. But more senior researchers (those who indicated more than fifteen years of experience in a field) offer further qualifications that directly implicate social relationships. For example, a director of research in statistics wrote that an expert is also "respected as a teacher; someone whose opinions are valued," and a public health policy adviser wrote, "Recommended, known, admired," while a communication studies professor wrote that an expert is also "humble." This latter notion of being humble is a refrain we heard from interview participants to which we will return, but it also connects with the idea of understanding the limits of one's knowledge (echoing Socrates, famously) while situating it within a social framework. To be humble is, in addition to a reflective position on one's own knowledge and knowledge limitations, a social comportment and allows one to build those relationships in teaching, mentoring, et cetera. But senior researchers, too, focus on what we might identify as more internal, individual characteristics of thought or knowing. A professor in education suggests that terms that apply to an expert include "robust, savvy, rich, broad, intricate, thick, rich," a professor of physics cites "wisdom," and a scientist in environmental physics adds "sometimes tunnel vision." Among survey responses, there are numerous instances of what we know, our experiences, but also how we situate our knowledge and its limits, as well as our social comportment. We will see, in this latter case, the importance of social comportment in how participants identify experts to collaborate with. However, before turning to those questions, we continue to develop some ideas about how participants define expertise and expert status.

Expertise and Problem-Solving

Interviewing experts provided further insight into the everyday operating definitions, or at least partially formed conceptions, of experts and expertise. For a rhetorical account, such everyday understandings uncover the complexity of these concepts as living, unfolding conceptions used in the negotiation of professional practice. When we asked participants "What does expertise mean to you?" many told us about the importance of knowledge and experience. Several, however, also discussed the important skill needed to deliberate on a problem or situation and to find the best available means by which to respond. These are important operations for which phronesis allows. Consider, for instance, several responses from participants who identify the social nature of knowledge production, the requirement of audience for understanding, and the idea of socially sanctioned ways of producing knowledge. Toni, when asked what expertise means, explained, "It's when I feel comfortable with the material that if someone asks me a question, even if I may not necessarily know the answer, I can think of a hypothesis that is reasonable to either myself or other individuals in the field, where I can, not necessarily just make a claim, but make a claim and be able to support that somehow." Hansaa similarly responds, "Expertise is basically understanding the subject matter completely, and understanding not only the topic itself but also perhaps different schools of thought on the topic, perhaps controversies within the topic area, and understanding what are some of the drivers and reasons behind some of those controversies, having a good understanding of what's been written and who it's been written by, and perhaps also kind of different approaches." Perhaps underscoring the attention to something beyond knowledge as episteme, or even techne, Jordan explains, "It's not about actually having knowledge so much as being able to understand how to find knowledge quickly. And I think more important than this, but slightly less formal than this, would be a sense in which when you encounter a new problem that doesn't have a solution yet, knowing essentially [from] what direction to attack it." In their responses, Toni, Hansaa, and Jordan all note the importance of having a capacity to situate a problem and find the correct, or at least reasonably well rationalized, approach and to solve the problem. It would be a mistake, however, to discount here knowledge as episteme or techne, as these too, in all three responses, are implicated. It is through an interaction among the different forms of knowing that individuals can perform at expert levels. Indeed, as Haru emphasizes, it is knowing-that which provides a fundamental basis on which experts are able to produce their work: "I think of it as being mastery of a topic and awareness of the greater body of work that's out there in the area, how it ties together, and what the important pieces are. An awareness and—well, I can't use the word expertise—some ability to explain and reproduce the most important elements of the research world." Situating knowledge, not merely amassing technical knowledge, however, requires the capacities for deliberation and judgment in order to identify elements of the situation, as Esmae notes, saying the expert must "identify what the problem is, develop a solution methodology, and come up with something that can be coherently implemented and effective," going on to note that they believe the capacity to complete this entire process is requisite for expert status. Shae states matters even more plainly, saying that expertise involves "having the right context to sort of know what's going on in your project or in your problem. So anyone can know any specific fact, but having a lot of related knowledge and understanding and intuition is how you become an expert to me," here implicating phronesis, episteme, and *nous* (i.e., intuitive thought). Pancha makes a similar point, saying of expertise that it is "in-depth knowledge basically of the art, and then an ability to approach problems that are outside of the ordinary and eventually come up with a solution," which Pancha suggests requires that one is "familiar with the procedural aspects of the field, but you also need to be able to approach new problems in that field." Attention to the problem-solving to which expertise seems to respond again invokes that refrain that constitutes phronesis as a practical wisdom or "good reason." But what of the moral aspects of the concept?

Theoretical and Practical Knowledge

Episteme, perhaps predictably, plays a vital role in how participants conceive of expertise, as does techne. This is a long-standing distinction between knowledge and skill (as techne) or practical knowledge (perhaps varieties of phronesis). Dana echoes what we hear from survey participants, saying that an expert "means someone who has the experience and knowledge in a specific area to be able to make decisions with infrequent mistakes." Notable, here, is the combination of experience and knowledge we heard earlier from survey participants, with the qualifier that experts should not be prone to mistakes. This is an important qualifier as it directly notes that knowledge and experience must cohere in some manner to improve experts' performance and decision-making, which is to say that a mechanism of training and practical reasoning must be involved as decision-making involves situational assessment. Rowan, in answering what expertise means to them, explains, "I think it's a combination of training and experience.... A real expert should have both a practical knowledge and a theoretical knowledge to be able to have the depth that's necessary"; Charlie similarly comments, "It means to be not only knowledgeable in a subject, but also be able to apply that knowledge into real-world applications"; and so, too, does Taylor, who says that experts "have a fairly in-depth knowledge of a particular topic, and they have demonstrated that, to me, usually in an applied as well as a theoretical way." Taylor goes on to caution that this might not always be the case as some experts do not conduct what we might call applied work, and so the heuristic is a general but not a universal one.

Application of knowledge is a complex idea itself and when used for expertizing can have different meanings across fields. Consider an example from Jamie, who says that expertise involves "knowing enough about something to sort of use it and think about things with it or through it," citing the example of "a critical methodology that you would learn from a different discipline. To become an expert in it, you have to do that sort of work of experiencing it and applying it once you sort of have, what you would say, a theoretical understanding, I guess. And a theoretical understanding in the sense that even theory can be applied." Application can be the particular skills, too, and Alex distinguishes between what they call technical and academic skills: "On the technical side I would attempt to do some real-time or relatively quick sort of assessment of specific skill sets to be able to do particular tasks. Do they know how to program in JavaScript? . . . It tends to be first very narrow and then from there, maybe I will step back to more sort of software—more general skills like, Are they a good problem solver? Are they organized? Do they stay on task? Do they keep to schedules? Can they articulate—communicate—to others what they know or don't know?"

Application is sometimes conceptualized as the ability to communicate findings to broader audiences or to instruct. "On the academic side," Alex explains, "I often tend to invert that order—and first and foremost I try to make some sort of assessment of their ability to communicate their ideas. So, How good of a writer are they? How good of a speaker are they? . . . I almost feel like I need to sort of generally assess their ability to communicate—and that then helps me better assess their level of sort of academic or scholarly expertise . . . —it's typically going to be revealed by one of those two modalities." Similarly, Marin explains that they see two facets of expertise: "One is knowing stuff and the other one is ability to communicate it." For Marin, both features are required, as is a moral imperative: "I always found that expertise comes with responsibility to share that knowledge." Communication of expert knowledge, however, is not simply a means of popularization, for instance. Angel provides further clarification on the importance of communication as a facet of expertise in a more complex sense than mere popularization: "It means an ability to explain why something ought to be done, explain why-related questions early, and then the how. So once you've figured out the why, you really need to have proficiency in all the techniques and the [inaudible] to implement the particular solution or plan or project."

Humility and Knowledge

What one knows, as our survey respondents reminded us, importantly involves knowledge of what one does not know, and our interview participants elaborate on this matter. Casey acknowledges the role of knowledge, specifically "more knowledge about a particular subject area than people who would claim to be nonexperts," as one criterion, but explains that "more than that, I think a particularly advanced degree of expertise is when people have a well-calibrated sense of what is known or true and what may not be known. And I think that's not often totally sort of monotonic with expertise." Casey provides an example in academia, where "junior graduate students may believe certain things to be known on the basis of a few papers published in the field's literature, but experts are able to put that into a broader context that may cast doubt on those kinds of findings or may inspire greater confidence in findings that actually would, on a superficial level, appear to have less evidence in support of them." Collins and Evans's (2007) work supports this position, where there is important social knowledge about the field required of experts for exactly this reason. Being expert in a field requires immersion in the field, in terms of the content and discourses, as well as the ongoing debates and actors. Another aspect of understanding a broader context involves awareness of cases that challenge our current understandings. Gail explains that it is attention to "all those little edge cases" that provides experts the capacity "to deal with the overall complexity."

As individuals begin to acknowledge their limitations they are able to fill in gaps in their knowledge, and an expert can be understood as "somebody who knows the history and all of the details of a specific field including both theory and practice and so somebody steeped in an understanding of a certain issue," Riley, an entomologist, explains. Riley adds that "then when you become an expert, that means you dig deeper into the details and you see all the dirty laundry. So that's an expert—[someone] who knows all the dirty laundry . . . but also knows where fields should be going." While somewhat lighthearted about the dirty laundry, this is a critical point Riley makes, that an expert is capable of recognizing problems or ongoing issues in a domain, but also can identify paths forward to help solve those problems. Mickey adds another dimension to this process of becoming expert, reminding us that coming to understand the field can involve a lot of moments where you do not understand the field or the problem. An expert, Mickey argues, is "somebody who's consistently failed a very narrow area of knowledge ... somebody who's failed from trying to approach that very narrow field of knowledge from various points of view." Such experience may be a vital component of a broader context that may, in some cases, require failure and learning from failure. In these failures, the "points of view" one gains allows for a complex formulation of the field.

Several participants also underscore aspects of becoming expert as, indeed, an ongoing process. Peyton defines expertise in such a way that the importance of recognizing one's limitations, and continually expanding one's knowledge, is necessary for problem-solving. Peyton first situates their understanding of expertise as having "developed a skill set and an awareness of a particular topic that others may not have," and, further elaborating, "that expertise is looked upon as a source of mentorship and guidance in those other people." Understanding the mentorship role of expertise has led Peyton to believe "there's also some responsibility that comes with that. As more and more people come to me seeking expertise on a topic, I think there's an incentive and a motivation to continue to keep up with the growing field, and also to recognize the limits of my own knowledge." Frankie similarly provides a definition that begins with something of a track record, but continues with some sense of ongoing efforts to become expert: "You have an established career in terms of whether ... you have publications, or if you are a physician, you have research projects, or you're a part of committees, and you're utilizing your expertise to push programs for quality improvement. That's sort of some track record. And, then also for me, I think expertise also—I am a big proponent of growth mindset—and, so, I look for—the expertise that I prefer to work with—are those that have that mentality of 'I'm still learning.'" One way to continue learning, Peyton explains, is understanding when to turn to others for necessary expertise, asking one's self, "Am going to become an expert in that particular field, or will I rely on others?" Time in the field is not merely a matter of acquisition and accumulation of episteme, but rather a more complex process of enculturation within a discourse community and an individual process of cultivating practical reasoning. Ashton also notes these features: "Expertise means that there is a degree of competency within a given field," adding, "I take that competency to mean both a theoretical breadth and kind of practical memory experience about different techniques being relevant in a given moment." Here the definition strongly echoes the idea of phronesis, providing an integration of episteme in "theoretical breadth," techne in "different techniques," and phronesis in "practical memory experience," memory being not merely recall but recollection.

Social Construction of Knowledge

Although the individual capacities that constitute expertise have been the focus for many participants, some responses focus on the social aspects of expertise, where the attribution of expertise or expert status is the foundational requirement. Quinn subscribes to an attributional model of expertise, explaining that the concept "is an attribution of authority by interlocular or audience member based on that member's presumptions about knowledge, abilities, or skills that you hold. . . . I think expertise, by definition, is a function of social role and attribution." Explaining how expertise is attribution, Quinn focuses on the situated role of expertise and experts: "Expertise is granted in a specific event for a specific purpose. Sometimes that granting is warranted based on desired outcomes, sometimes that granting is a mistake, right, sometimes that granting is a category error, but it still functions socially as expertise in that moment even if it didn't lead to the desired outcomes." Roshan provided a concrete example of how this might work, as quoted above: "I can give you my understanding of the legal thresholds in place. Every state has the opportunity to define it differently in the US. But it really means knowledge a layperson wouldn't know about a topic. So simply having a college degree could technically qualify you to be an expert in a field." Understanding expertise as a contextual event is useful beyond an attribution model alone, because expertise is often specialized. Although expertise is not always specialized to the degree it can be deemed interchangeable with specializations, such as an area of research, there is normally some form of focus. For example, in this study, expertise largely overlapped with specific research areas or industry roles. However, as is the case with Harper's (1992) study of a generalist expert, Willie, who ran a small shop, sometimes an expert can adapt a wide range of materials and tools to respond to a particular problem. Expertise, this is to say, is not always specialized in the sense of academic specialization or professional specialization. In this way, when we think about the contextual nature of expertise—the researcher trying to solve a problem in a niche area, or Willie working in his shop—what becomes important to understand is the problem-based nature of the work. Demi argues as much: "I would define expertise [as] your ability to say if you're going to do a task, whatever it is—it can be theoretical, it can be technical—that you are going to be able to carry that task to its end. So expertise is more of the main specific thing, rather than something that defines an individual just as a whole. So you can be an expert in or an expert at but being an expert I don't really believe in that anymore."

Defining Expertise Across Disciplines

When experts were asked about where expertise has common or differing definitions across disciplines, the goal was to understand how they might theorize why the heuristics they use in their own disciplines to assess experts may or may not be transferable to other disciplines. To some degree, this comes down to a matter of scale in assessing expertise. Technical skills one might associate with typical experts in their own specialized area are unlikely to transfer to others directly.

However, as with the definition of expertise, the participants interviewed had developed somewhat broader understandings of the concept. "It's certainly not a generalized notion at all," explains Mickey, who works in mathematical physics. "The definition itself of expertise varies vastly over disciplines. And not only have I observed this in my own field. Whenever you move from one specific field of theoretical physics to another, the notion of who an expert is and what makes somebody an expert changes drastically." Marin draws the distinction somewhat more broadly, between those disciplines, normally in STEM, that require a narrow focus for knowledge production in contrast to other disciplines, normally those in the humanities and social sciences, that require broader contexts. Conceding they are "operating in stereotypes," Marin explains, "I do think that in more technical fields, the deeper and the more narrow the expertise gets, the more glorified it is. . . . And there are people who tend to go very, very deep but fairly narrow, and there are people who tend to go broader but by necessity, more shallow. And I think in technical disciplines what is rewarded and admired most is when people go deep rather than broad." In the social sciences, Marin continues, "people do go deeper, but that deeper tends to be limited because you need a broader context." Providing an example, Marin suggests that "whereas in if you're in physics, for example, you can just devote your life to one particular type of black hole. . . . Whereas in most social disciplines, you need a broader context to it because you can't quite speak about people without actually knowing a lot about what they do." Both Mickey and Marin are suggesting key features of expertise, including disciplinary knowledge. When someone is considered an expert is another important and disciplinarily or domain-constrained matter. Frankie explains that "some people maybe say . . . you're an expert after five years," using the field of medicine as an example, but adding a contrasting example, saying "if you're in academia, you're not an expert until you're at least a tenured professor with ten publications and the [federal] grant."

The forms of socializing in each field, as well as the nature of the problems, demand different attributes of their experts—for example, cognitive and kinetic dexterity versus considerably accumulated knowledge of prior research. Although there may be differences in how an expert is evaluated, Jordan, an expert in decision theory, explains that there are some commonalities to be found in the goal or outcome of the activities. Jordan uses the example of a mathematician versus a biologist, explaining that "understanding how to solve a new math problem versus understanding how to solve a biology problem, one of them might be a lot more physical skill working in a lab, one of them might be more conceptual." Continuing to clarify that, although there are distinctions, there is a commonality of purpose; Jordan says, "I think it's essentially different services in the pursuit of the same goal." Here the situational nature, the social conditions, will shape one's expertise, and likely notions of ethics and prudence are also notable.

One recurring theme that stood out, especially from my own vantage point as a rhetorical scholar firmly situated in the humanities, was how often STEM researchers expressed uncertainty about assessing expertise in the humanities and arts. Harper, an expert in statistics, believes there is considerable "uniformity" in definitions of expertise, and explains they are "accustomed to seeing research in physics, research in chemistry, and research in biology, all with the same underpinnings." Harper continues, however, to say that such uniformity in STEM fields may not be as easily discernable in the humanities or arts. "It makes me a little awkward when I meet somebody who's an expert in literature or music," Harper notes, adding that even questions about areas of expertise become challenging, such as "Do they know music theory or just that they're a great performer?" Harper isn't alone. Siobhan, when asked if definitions of expertise are fairly consistent across fields, agreed: "Yeah. I think it's pretty standard," adding the qualification for a purported expert in the arts: "I don't know what their expertise would consist of, but for your more standard academic fields where you're not producing a creative product at the end, I guess, then I think yes." Haru, an expert in computer graphics, similarly comments, "I'll suggest that my opinion on what constitutes expertise—that that's a useful point of view for a lot of STEM fields for sure because those disciplines are driven by publication, by conferences, by individual researchers or teams having breakthroughs and disseminating that into the community. But I'm not sure that's the way it plays out in all disciplines." "I mean," Haru elaborates, "if I'm thinking about humanities, or fine art, or dance, or creative writing, the breakthroughs don't come through scholarly publication dissemination that way."[2] Ultimately, Haru says, "I'm not sure what I would look for as expertise there." Angel, who does identify a difference in expertise uses an example of how different domains approach problems: "Let's say something like physics, and mathematics, and geomatics—there is a theorem, and you can either prove the theorem to be right or wrong.... For the expertise of people who are in a more arts-related field, that is different than the expertise of people in axiomatic science." Regardless of whether you agree with the articulated distinctions between STEM and humanities, social science, and arts, there is some confusion about how to assess experts across domains. This, more anecdotally, has been my experience, too, working in multidisciplinary teams. Important in these articulations is that there is some sense that there are differences in how different fields make knowledge, and that such variation does not delegitimize one field or another, but it does make working across disciplines challenging. Further, these interviews provide a window into the stories we tell about how our disciplines make knowledge and how our discipline is distinct from others. But, in fact, it would seem the matter is even more involved than that, and Julio, who works in software development, identifies some of those complexities. Challenging the question we posed, "Do you think expertise means different things across disciplines or is it a fairly standard notion?" Julio wanted to "twist" our question and answer a somewhat different version. "Expertise," Julio tells us, "means different things to different people because there's different people in all disciplines." "But, to answer your specific question," Julio continues, "I would agree that there's a certain level of knowledge, a certain bar," and this bar is true for "mechanical engineering," "sculpting," et cetera.

Assessing Expertise

In a Likert-based survey question, we asked participants to express their confidence in evaluating others' expertise. The results echo the previous discussion. There is a notable trend resulting from these questions where, when asked about one's own specialization or subdiscipline, a high degree of confidence is expressed in evaluating the expertise of others. Even moving into one's own field, this confidence somewhat diminishes, with "more" confidence or "moderate" confidence rather than a "very" confident ranking. Moving even further away from one's own area, but still within the broad framework of either STEM or humanities and social sciences, confidence again diminishes. Finally, when trying to evaluate expertise far outside one's domain, such as a STEM-based researcher trying to evaluate the expertise of a humanities or social sciences-based researcher, the "not" or "only slightly" confident evaluations become much more common.[3] Assessing expertise is a complicated matter and, as we learned from the participants we interviewed, there are a number of strategies one might use to do it.

Feeling Out Expert Status

We asked interview participants, "How do you tell if someone is an expert in an area where you're not an expert?" Riley explains their process of identifying who is an expert: "I will go and look at people's articles and go to Google Scholar or go to Web of Science and see who's been quoting them and see how they're networked into the system and get a little bit of a sense from that if I'm making the right judgment call." Then, Riley, in a theme seen throughout processes of identifying potential expert collaborators, turns to their own social network, saying they continue the process by "talking to colleagues." In another common thread, Riley also sits down to speak with a potential collaborator: "Email and phone calls are just not the same as sitting there with somebody in a room and, I know those are subjective judgments, but when you sit in a room with somebody and talk to them about issues you can see a lot about how they answer questions and how they respond to things. Whether they're there just to impress or whether they're there to be inquisitive." Indeed, this final statement draws our attention to questions of eunoia (goodwill), a notion that is intimately connected to the moral dimensions of expertise. In a research effort, attempts to impress (a vice, perhaps, of pride) are critically evaluated. Indeed, as we see throughout the participant responses, researchers are interested in other researchers, other experts, who draw attention not to themselves or their research, or even their own research paradigm, but rather to the problem at hand.

Julio, recalling what we might hear as an Aristotelian refrain, explains that the process of becoming an expert is one of "trial and error," adding, "I think you get better as you get older. And I don't mean in your particular field. I just mean in general because you see certain behaviors." In almost a virtue model, Julio notes some of the "red flags" they look for, saying that "if I'm dealing with a person who is in a matter of a day or two reminding me of other projects they've been on where they did this great big amazing thing—they've told me the third or fourth time—I get a feeling that they're—actually, [it's] a lack of confidence." This lack of confidence motivates someone to retell stories of their success, Julio posits, is because they are trying to persuade, "telling me that so that I think they're amazing." As a "red flag" this causes Julio to listen to the purported expert more carefully to discern what motives might be behind the tales. Another red flag, one that applies across both the "arts and technological fields," that Julio watches for is when people are incapable of explaining their reasoning or providing a clear explanation to others. "I will ask you a question and if you just kind of give me a rote answer or if you try to techno-babble me"—adding that one can also "techno-babble in arts"—Julio then will "ask more questions, and if you cannot explain to me in simple terms something we're talking about, then you probably don't know what you're talking about."[4] Experts are not merely storehouses of knowledge, and Julio reminds us of the social relationship experts enter by virtue of their designation as people who can apply and share their specialized knowledge, not simply retain it. Similarly, Kendall explains that they assess expertise in some common terms, such as "being able to cite relevant past activities and being able to apply those results to the problem we're working on." However, Kendall employs some of the same techniques and principals as Julio by way of questioning and looking for explanations as well as looking for someone to situate research in the scholarly literature, saying, "Even if I couldn't have come up with that citation myself, I can at least read it [and see if it is] applicable or not. . . . I'm not the expert. I can't tell you if they're the most applicable things like an expert could, but the fact that they're applicable is pretty good evidence that the person is certainly more expert than me."

Research Record and Affiliation

Dallas, an expert in statistics, often first encounters a potential collaborator through written work, as many of our participants indicate they do. However, for Dallas, rather than reading a peer-reviewed publication, it is often through researchers sending a project pitch that they first encounter someone textually. For Dallas, the pitches are potential collaborators "describing something that they would like to do in a couple of pages. So they're trying to describe typically something that might be actually fairly technical and fairly disciplinary in writing." Further complicating matters, in many cases the researchers are not experts in Dallas's field but have some expert knowledge that can contribute to the problems on which Dallas's lab works. In this case, assessing someone's expertise can be challenging. There are those researchers who submit well-written project descriptions, clearly describing what they have proposed, including challenging problems they hope to address. "I don't necessarily understand the details of how they would go about doing the problems," Dallas explains, "but they are able to articulate what they do in the context that *I'm interested in* and articulate for me what's interesting and hard." Dallas explains that in their assessment of expertise, they believe that "there is some requirement that the people on the other end [the experts] have to be able to *give me hooks and cues* and talk about things in a way that gives me some entry to start to understand what they're doing." For example, Dallas has expertise in statistics, so they are able to understand poorly written prose if the statistical descriptions allow them to work backward to understand the problem: "So for particular things within my discipline, I am able to put on my disciplinary thinking cap and actually think in math and symbols and try to understand what they're doing." In contrast, "for people in other disciplines, if someone is talking to me that way about sociology or computer science, I don't have a lot of ways to figure out whether they're talking nonsense or not." In multidisciplinary teams, then, an indicator for Dallas of whether someone is an expert relies on their "clarity of thought and an ability to describe this thing that you are expert about."

Avery explains that, by and large, when they are looking for someone with whom to collaborate, they are "not really as concerned with the idea of them being the expert" because most of the experts they would be collaborating come their university. This is to suggest that trust is built into the mechanisms that help vet would-be experts before they become part of the university community. Avery explains their school is a "good research university, and so what happens is anyone I meet on campus has probably passed some bar of expertise." Jamie also invokes similar ideas of social trust, explaining that the relationship is "going from a sort of an *honor system* on the basis of them [a potential expert collaborator] being honest." Instead of relying solely on the high bar set by colleagues, Avery looks "to the people that I get along with. . . . People that I can talk with easily. People that I can ask more questions if I don't understand something." Avery's concern with finding someone they can work with is also reflected in the survey responses, as we will see below. However, other participants in interviews identified the importance of being able to work as a team not just with people with whom they get along, but those with whom they might productively collaborate. Caden explains, "Humility, I think, is one factor that typically shows you [who is an expert]"; someone with expertise will be able to rearticulate information when communicating with a nonexpert in a manner that is "well-rounded . . . rather than always bringing the topic back to one special area of focus." Caden adds that they look for someone "who seems interested in the topic, as well, is a sign of expertise that I would usually look for."

When we discuss the idea of a track record it is perhaps too easy to have that mean publications; in fact, our participants tell us that previous work demonstrates a complex constellation of capacities as well as social evaluations and norms related to trust. Jamie, for instance, says, "One easy way that is sort of empirical in that you've seen them working on these projects, and they've completed them, and you can sort of tell on the basis of the result." Results do not simply mean peer-reviewed publications. Rather, Jamie believes that you can assess on the basis of, for example, "using a program and having it work." The ability to broadly situate one's work, Jamie notes, is partly based on "how your work on this particular project matters in a broader sense. So how is it that it connects to other projects you can be involved in," adding that it is also critical to understand the "theoretical implications" of one's work. "A high level of expertise," Jamie continues, "would go hand in hand with that sort of ability to apply more widely." Caden provides another prospective on how we might assess someone's record of accomplishment: through their capacity to understand the complexities of projects. "I typically find that the most expert people have a lot of understanding about the difficulty of the things that they're interested in," Caden says, "and they don't necessarily try to portray it as an easy thing." Caden argues for why this measure is effective, saying that focusing on the difficulty of problems "seems counterintuitive, I know, because experts should find these things easy. But at the same time, I think expertise requires a more kind of well-rounded understanding of a topic and being able to say, 'This is a complex thing.' Because anyone can be an expert in very simple tasks."

In addition to assessing experts' records of accomplishment in publishing and research, or their affiliations, one might try to themselves understand the disciplinary conversation of the individuals they are assessing. These strategies, however, rely on trust within networks and interpersonal communication. Marin attempts to assess someone's expertise relative to the community of experts of which a person purports to be a member. "I tend to check if something I heard from my expert is something that the discipline or the area would generally agree with," Marin explains, cautioning that "this is not to say that people who tend to be against the grain in any particular discipline are wrong necessarily, but there is a point at which if you're just one voice, and there isn't much evidence supporting those particular views that there clearly is something off there." Such evaluations are a challenge, however. Haru asks, "If you really are not an expert, then what can you do?" in terms of evaluating someone's expertise, putting the evaluation down to, in part, "instinct." In making such instinctual judgments, one is, in fact, using several heuristics for social trust. Haru notes insightful questions to ask of how we view a purported expert: "How do they comport themselves? What's their attitude toward their research?" Interpersonal evaluation becomes crucial, and Haru says that "the best method for me is I want to have a conversation with [potential expert collaborators]. I want to sit down and talk with them. And I want to come away feeling energized, excited, feeling like there are some new ideas that are worth exploring." Questions of trust and interpersonal relationships were also discussed explicitly by our participants, a subject to which we will now turn.

Trust, Ethics, and Reliability

Survey participants were also asked, "What other factors do you believe should be considered when evaluating someone's status as expert or their expertise?" One data analyst wrote, "I can definitely say that degrees and the institution from which they are granted does not correspond to someone whom I would trust. They may have expertise, but I would not necessarily trust them to be ethical." Such a statement powerfully illustrates the relationship between trust, expertise, and expert status. If we cannot trust experts, even among collaborators and colleagues, the relevance of experts might rightly be put into question. A statistics professor illustrates this point, writing of other factors that are important to evaluating expert status or expertise: "Ethics, rigor. So-called experts that lack those can be dangerous in a number of ways, especially when recognized by their peers who may be blind to the lack of rigor." A biology professor puts matters even more plainly: "Just because someone is an expert, it doesn't mean that they aren't also an asshole."[5] They continue, explaining that "if you want to work across fields (or even within a field), you need to identify people who are good, communicative, reliable team players as well as experts. An expert who is unreliable or flaky is a total waste of time."

The social function of experts even within specific fields or multidisciplinary teams is important, as it is when experts interact with publics. Themes surrounding trust, ethical behavior, and sociability gesture to the characteristics of moral knowledge discussed in previous chapters. It is difficult, then, to imagine that experts might normally operate without regard to these forms of knowledge. Although the tortured, abrasive genius or expert may be a popular trope of prime-time television, the participants we interviewed indicate that real experts go to some effort to ensure that they avoid such characters. Indeed, a research director with a specialization in statistics wrote, "Ideally, we'd have some way to measure levelheadedness or reliability. We usually don't." One approach, however far removed from a reliable measure, is to talk to prospective collaborators. "How well they work with others," a senior architect answers, "this does not reflect on their ability or knowledge of their field, but helps determine how much effort it would take to get the information needed for the project. An unhelpful super-genius is less useful than a well-spoken genius." To determine if someone would be helpful or unhelpful, the architect explains that they "would also have at least one semi-social 'interview'—perhaps a lunch or a stroll through a park—talking with a person outside of the normal work environment can be insightful." An entomology professor echoes these sentiments: "Someone can be an expert based on knowledge . . . but in terms of evaluating their status as an expert who can be helpful, I evaluate their trustworthiness and reliability." However, assessing these qualities remains a challenge.

Assessment of expertise and expert status relies on both social heuristics as well as forms of critical evaluation. An economics professor explains that their criteria include "publication record, field, gut feeling." Gut feeling invokes a range of tacit knowledge related to social trust that we cannot explore at length here but is crucial for understanding our social judgments of others and how it might perpetuate biases. A research scientist in human factors, drawing on that kind of trained skepticism of a researcher, implores, "Don't just look at one source to 'determine' expertise (such as a CV or résumé). Oftentimes people are only experts on paper, and it takes some digging to understand the depth of expertise." They provide an example: "If someone says they are a director of a program, it is worth checking what the program entails. Are they a director in name only or do they supervise? What's the track record of the program? What collaborative work have they done within that program?"

Another method of assessment is to look at the ability of the would-be expert to focus on the problem or situation in such a way that it takes priority over what we could perhaps call one's ego. A neuroscientist explains their criteria for evaluation: "If they express too much or too little [confidence], I am suspect. Usually experts are more comfortable expressing what they do not know, or what is not known," again invoking a sense of humility among experts. A data scientist explains, "The context of their knowledge is to me what sets an expert apart. How well can they carry over what they know to a new situation?" A comment echoing this statement comes from a statistics professor who wrote that "understanding the context in which their knowledge resides" is important for an expert.

Identifying Collaborators

In our survey, we asked, "How do you identify researchers outside of your discipline you would like to (or need to) collaborate with on a large project?"[6] "Recommendations from colleagues," a research scientist in computer science responded, a common refrain among experts. In biology, a professor wrote, "Word of mouth— I'd ask around in my professional network to see if anyone had a colleague who has expertise in this area." Extending one's professional network in a research university may involve more than one's own, established circles. Latent networks, too, are important. A postdoctoral scholar in chemistry provides an example of such latent networks, telling us they would identify potential collaborators "through the grants office, or personal connections/recommendations." Continuing the theme, an architect wrote, "I would first reach out to my 'trusted' network—individuals I have worked [with] successfully in the past," echoing other participants about the importance of their networks of colleagues to provide some initial contact. However, the architect continues: "Note in this case 'successful' does not relate to whether or not the project achieved its goals, but rather to how well we got along and how well the person did their job." This statement recalls the idea of phronesis in expertise, where the identification of the situation, its parameters, and the available and appropriate means of response, through deliberation and judgment, do not necessarily result in success all the time, but rather promise a well-reasoned, practically wise approach. Continuing to elaborate, the architect explains, "Next would be people who I may only know socially whose opinion I *trust*. Credentials are easy enough to determine through Google and LinkedIn, and a well-structured set of interview questions prepared by a trusted expert in a field will help weed out any posers," which invokes the idea of ethos in expert status. In a quite different field, communication studies, a professor provides a similar account, writing that they would assess a prospective collaborator's "publication record in the relevant field—impact measures plus quality to the extent I can decipher it; their participation in related projects; their reputation in communities or with people I trust; their willingness to be up front about the limitations of their knowledge." A physics professor describes using their network, too: "I would first contact someone I knew as a reliable source within that discipline. If I knew nobody like that, I would ask colleagues in my own discipline whom I trusted if they could identify someone with whom I might collaborate." Credentials, we learn from these participants, are not in themselves enough to be persuasive. Rather, matters of trust are critical. Returning to the architect, matters are put plainly in this regard: "It is the character of the person and how they work with others that is as important as the person's knowledge."

Track record, which is a key component of assessing expertise, was mentioned by numerous participants as a means to identify potential collaborators. An entomology professor wrote, "I look up their publications," but elaborates that they engage in further assessment: "I meet with them in person to judge their general knowledge, ability to think broadly, and to understand their past collaborations." Another influential factor the entomology professor noted was a matter of affect: "I find that people also self-select by enthusiasm for the project and by actively getting on board." Similarly, a research scientist in human factors wrote that they identify "history of interdisciplinary research, track record of successes and failures," but, again, continued by identifying more personal attributions of potential collaborators, saying they also look for "willingness to acknowledge limits of own expertise, [being] upfront about resource capacity (e.g., time, effort)." What appears to be at work in these comments is not merely identification of credentials, although these operate as an important preliminary indicator of expertise, but also the capacities for thinking collaboratively. A scientist working in neuroscience explains that they identify possible collaborators "by judging how they regard problems in their own field or closely related fields." An emerita professor in computer science summarizes another key attribute that seems to underlie much of this discussion: "respect for collaborative interests." Participants elucidate that they are not simply interested in the technical expertise one brings to a problem but the capacity for identifying, deliberating on, and solving problems within a team, which requires acknowledging the limitations of one's own knowledge and perspectives and integrating with others to improve problem-solving approaches.

In interviews, a similar question was posed to participants: "How do you identify experts to collaborate with from outside of your discipline or a closely related area?" One of the most commonly cited methods of evaluation was drawing on one's social network to identify possible collaborators or to vet these prospective team members. Some begin with a web search, finding what information they can in online faculty profiles, for example, and consulting the peer-reviewed publications listed. Although credibility is often initially established through engagement with written materials, such as publications and or a curriculum vitae, the assessment soon turns to questions of how would-be collaborators function in teams, how successful were their past projects, or how well they get on with others. Caden provides a fairly typical overview of processes we learned researchers often use to identify prospective collaborators: "I would look up some credentials. I would consult their publication record, if it was a relevant field where they published. I would consult their CV. I would, if possible, talk to the person." Similarly, Dallas explains that when beginning their lab, the first step was to "Google around and read CVs" to figure out who might be working on problems similar to the problems the lab wanted to pursue. "You can certainly tell if someone is research active" using this method, Dallas notes, but "that's sort of a first screen." After the initial screening, "then you contact the person and you describe what you are doing and then there's a process of writing and conversation to try to assess." For Sam, the real possibility of a collaboration "almost always starts with a conversation. It's 'Who do you know who's done this?' or 'Who has shown an interest in that field on what I'm doing?' So it starts with a conversation and just people's reputation."

A key strategy for academic experts is to understand the publication record of their potential collaborator, even early on in these processes of searching online or identifying prospective collaborators through reputational networks. Assessment of potential collaborators through social networks, and efforts to establish credibility through publication records, are indeed an interconnected effort. "I would normally go on Google Scholar and search for relevant terms or go to university websites and go see if they have a department that's in that area and see if they have anybody that's relevant," Siobhan explains, adding, "The third way would be to ask people, 'Do you know anybody who's an expert in this particular area?'" In evaluating publications, Gail states that "in the sciences, a person's publication record is often a good indication of whether or not they know what they're doing and how good they are at it in terms of number of papers, quality of journals those papers are published in, as well as where they are on that author list. If they're always buried in the middle, then possibly they've just been playing a support role." Similarly, Hansaa says, "In academia, you would look at their academic record. So you would look at the post they hold, the university they hold it at, the publications they've had, the quality of the journals they've been published in, who they're working with. So again, the *quality of their collaborators*."

Gail provides further insight into how this might be done: "Even if you don't understand the paper itself, you can get a sense for how much of an expert this person is if they're successful in publishing because you can presume that those papers have been reviewed by people who are experts, who do share that domain-specific expertise." Returning to the idea of having conversations with individuals to assess their expertise, their publication record and understanding of the field, including emerging and perhaps yet-unpublished research, can be a key indicator. Experts, Casey believes, will "usually be able to cast some additional context around a finding. And particularly when that has to do with information that may not be published, that's a really good sign that they're an expert." Nuance in the discussion is what Casey looks for in an expert answer: "Another thing they will often do is point at a kind of detailed aspect of the methods to either comment on them being unusual, or particularly powerful, or potentially weak, or so forth." Shae adds another dimension of how such conversations are assessed—that is, instead of listening to what the expert knows (that perhaps is not documented in the literature), Shae listens for what is missing. "I pay a lot of attention to *how people talk* in absolutes or not," observes Shae, saying that an expert will qualify statements along the lines of "Oh, yeah. There's all these special cases, but we're not going to talk about that right now." One of the benefits to these approaches is they do not require the listener be an expert in the field discussed and can therefore be a useful heuristic for experts working in multidisciplinary teams.[7]

It is not only publications that researchers look to in order to identify possible collaborators, but more generally a researcher's record in completed projects and how well experts can describe those projects. "I look at their past projects," Toni says. "Not necessarily their publications, but what projects they have worked on." Blair also highlights the importance of projects: "I pay attention to *the experience the person has had* because I think experience definitely contributes to expertise. I'd pay attention to knowledge and the information they've shared. Those are definitely areas where, based on if their confidence about the work that they've done, and the experiences they've had in their field, helps me get closer to knowing that they are the expert." Pancha uses a similar technique in industry: "If you've worked on a project that is related to an area, then you might be considered an expert in that area. So it's essentially just tracking if you've had some real exposure to a given subject beyond looking at the textbook." Toni explains one rationale for such an approach; in addition to noting whether someone has a publication, they also "look more at the methods that they [potential collaborators] used, and the tools that they use. So for me, it's less about the end product of what they did, but what they used to get to that end product." Although reviewing someone's credentials and past publications is important, this is often seen as only the first step. Especially important here is that effective conversation and communication reveals the relative capacities of experts to identify the limits of their knowledge, their abilities to work with others on a problem, and also their interpersonal agreeability or comportment.

Angel explains that, when looking for a collaborator, they will reach out to their social circle, even asking about them on sites where they have an established network and may be part of groups with relevant expertise or connections to other experts. Through these networks, Angel says, one might find a "similar person who not only has the expertise and resources, but also *their personality sort of goes for your team.*" When asking their network or "friends of friends," Gail asks "Do people speak highly of this person or not?," noting this can be problematic for a number of reasons, and so one ought to also ask, "Are they really making those decisions based on the person's expertise? Or are they making those decisions based on some *personal qualities or likability?* Someone who's likable is not necessarily good at what they do." For Riley, "one of the things you look for is somebody who has that expertise but is a bit *humble,* and also interested and has the capacity to listen to the other disciplines." Explaining why this is important when identifying would-be collaborators, Riley says that when "somebody comes off as really expert in their field, but dismissive of other fields, that becomes problematic." With extensive multidisciplinary experience to draw from, Riley continues, "you go sometimes and you look at all of this and people are saying things that are not in your worldview. They don't seem acceptable to you because that's not the way you look at things. And then you look and you find out, 'Gee, but these are really smart people.'" Summarizing why capacities to understand others' worldviews are essential, Riley argues, "That's really important to look at that from the sense of, 'Wait a minute, how did that person come about their perspective if it's not one that can come to you through the lens that you use?' And it's important to *see in the people that you're talking to, whether they have the capacity to look through other lenses.*"

What Professional Experts Reveal

Among the experts who responded in both surveys and interviews, a few common strategies were used to identify potential collaborators, including looking at someone's track record in terms of publications and past collaborations, deploying a social network of colleagues or through their home institution to identify individuals, as well as sitting down with these would-be collaborators to make judgments about personality, intention, and, expertise. When possible, experts take considerable time and energy to vet those with whom they might collaborate. With the range of fields, levels of expertise, and experience in multidisciplinary settings among our participants, it is perhaps surprising how common the strategies of assessing expertise seemed to be, much like the commonalities among definitions of expertise. It is not entirely surprising given that most participants have had extensive academic training, where socialization helps shape what we assess and value in experts. However, it is also apparent from our participants that the operating definitions of expertise as well as the strategies for assessing others' expertise or expert status are also complex, not simply relying on technical measures, for example.

When Collins and Evans (2007) discuss meta-expertises, they explore how we are able to make judgments about expert claims or expert status. Their "Periodic Table of Expertises" provides two types of meta-expertises: transmuted and non-transmuted. Transmuted expertises do not discriminate on a technical criterion, but rather focus on social expertises to make judgments (Collins and Weinel 2011). Non-transmuted expertises are grounded in technical discourses and include technical connoisseurship, downward discrimination, and referred expertise. There are also, in addition to meta-expertises, meta-criteria that are used to assess experts and that allow us to judge expert status and expert claims. Meta-criteria include credentials, experience, and professional or research track record. But phronesis operates somewhere between transmuted and non-transmuted, so it is not surprising an analogue does not appear in their model.

To have phronesis as a central constituent of expertise or expert knowing of the expert is to re-center the audience for expertise and expert knowledge. The audience, however, is vastly more complicated than the singular construction suggests, and phronesis, with its ethical core, helps experts/rhetors understand their duty to an audience and relationships with an audience in a particular situation. Publics can be rendered vulnerable by expertise, and expert status (excused of any ethical responsibility) is an oxymoron in that the supposedly authoritative knowledge experts would offer is not the best knowledge for the situation. For example, nurses who spread anti-vaccination messages might very well have experience as health professionals, and even professional status that would suggest they have authoritative knowledge, but they are not generally experts in epidemiology, vaccine development and distribution, and public health. Deploying expert status as a persuasive technique here is distinct from being an expert in the topic at hand. In this example, the neglecting of their moral duty to their audience as an expert can cause physical harm. In contrast, sometimes there is cause to doubt someone who seems to be a credible scientific expert even when they share factual information, but information divorced from the real-world consequences of their audience. We see examples of this among citizen science projects that emerge in response to inequitable distributions of risk, such as toxic waste siting (see, for example, Pezzullo 2001). In the next chapter, I look to the case of citizen science to further explore and complicate the challenges raised in these discussions with professional experts. A rhetorical understanding of expertise requires looking to what relational attributes underlie expertise's nature and function. Citizen science offers a fruitful case to look at such relational attributes as it does not merely rehearse expert and nonexpert binaries, and that is the subject of the next chapter.

5

Citizen Scientists on Expertise

Imagine yourself at a riverbank in the summertime, water rushing by as multiple refrains sung by a multitude of insects merge with those of the birds perched nearby among the trees. On a normal hike, one might simply take in the scene, enjoying the abundance, but citizen scientists also have a mission. In this example, imagine citizen scientists first identifying plants, which they know to be the most likely habitat for a species of beetle they are monitoring.[1] As they scour the riverbank, they are part of a larger enterprise of scientific study, a phenomenon called citizen science. As noted above, citizen science describes everyday people becoming involved with scientific research efforts. There are several ways that individuals participate in scientific research in this model, from assisting with data collection and analysis to the very design of research programs. Normally, this practice is described in the research literature on citizen science in two distinguishable forms. The first is where a scientist establishes a project, for example, trying to figure out what kinds of galaxies are in pictures, as in the Galaxy Zoo project. In this project, citizen scientists work online, from their own computers, to help scientists analyze data by sorting distinct kinds of galaxies.[2] In another form, "grassroots" citizen science is designed and carried out without professional oversight.[3] Examples of this would be the citizen science groups that emerged following the nuclear disaster at Fukushima Daiichi in 2011 who strived to measure contamination of land, water, and food.[4]

Grassroots citizen science is especially interesting for questions of expertise. Citizen science projects can indicate inadequacies of expert responses. The growth of citizen science in response to disaster, risk, and hazard marks, to a degree, the failure of experts to recognize the importance of rhetorical character, of ethos, in their relationship with publics. Such forms of citizen science can also be traced to important social and environmental justice work (see, for example, Bullard's [2018] important study of race, class, and grassroots environmental work). Kimura's (2016) study of community-based monitoring for food safety after the

2011 Fukushima Daiichi nuclear disaster demonstrates the emergences of citizen science where there was a lack of appropriate expert response. Kimura shows how gender, scientism, and neoliberalism affected community-based monitoring organizations' efforts for testing food contamination. Many of the citizen scientists were women concerned with food safety, and through gender politics and policing they were dismissed as "radiation brain moms" who were not engaged in science—and, moreover, distinctly *lacked* scientific knowledge—but rather were participating in *fūhyōhigai* (harmful rumor). Kimura and Kinchy (2019) provide a broader account of such situations, explaining how the virtues of citizen science are sometimes oversold, while the vices, or at least the dangers, are overlooked (see also Kimura and Kinchy 2016). Citizens who participate seriously in deliberative processes, such as the women who ran community-based food contamination monitoring organizations after Fukushima, often have an external urgency, such as immediate environmental, health, or safety threats that motivate and sustain their engagement. Engagement in these cases is paired with considerable experience, including local knowledges. Moreover, the move to bridge citizen scientists' knowledge and experience with the forms of sanctioned scientific discourse—including the ways knowledge is articulated, validated, and debated in science—shapes a notable form of expertise that might shed new light on how we understand the concept. Citizen scientists might learn the languages of science—and often policy as well as other important bureaucratic discourses—to appeal to professional and institutional experts for change. Although these cases are not always successful, it is the effort to reconfigure expert-public relationships that is especially interesting to the question of expertise. Indeed, expert status is often a highly protected status. The professional and social debate about the relative status of experts demonstrates the important rhetorical function of such questions: expertise is meant to be somewhat exclusive. Because expertise is exclusive, experts have a moral duty to share their knowledge when it comes to bear on a situation that, without such insights, cannot be solved.

We interviewed over twenty citizen scientists, who volunteered from around the world and represent a range of backgrounds. Citizen scientists note the importance of theoretical knowledge, professionalization, and so on, that we also heard from professionals. However, citizen scientists also provided rich insights into conceptions of expertise that are not coterminous with what the professionals might offer. Interesting to this study, several participants noted affective dimensions required for lifelong learning. For example, Max tells us of the "joy of sharing information for something that I love"; Bailey speaks of "having natural curiosity, innate curiosity," as key to cultivating expertise; and others spoke of the importance of practical knowledge gained through experience, as Sean notes by saying they gained expertise in part through training but also by "having done it for twenty years. You just gain experience." Experience as central to building expertise is also built into discussions of the importance of local knowledge. River tells us that a citizen scientist "might observe something that scientists can't possibly observe because they're not there." Immersion within a particular context, knowledge of its edge cases are all important especially when we consider complex systems, and this kind of experience might well be central to becoming expert. Here again is an issue of knowing how to structure knowledge surrounding a problem in a manner that is both theoretically informed and also practically knowledgeable, inclusive of a practical moral knowledge. Citizen scientists note the importance of understanding social structures, too. Reese explains that it is important "knowing the lay of the land or knowing who [others are] connected to"; and Ashley likewise tells us of the importance of understanding how knowledge is structured through their work in libraries, which also signals importance social conventions surrounding the structuring of knowledge.

Interviewing a range of participants involved in citizen science efforts provides another view on expertise. This chapter explores how participants conceptualize expertise while assessing experts and how they believe they themselves became experts. Understanding how citizen scientists move from amateurs through novices to experts can help develop innovative approaches to thinking about democratic engagement in technical and scientific subjects. It can do this by dispelling simplistic notions that citizens are somehow incapable of serious engagement with experts for democratic processes. Further, such a reconfiguring of expertise also, I argue, illustrates critical expert capacities we might mark with phronesis as lacking in our contemporary popular rendering of an expert. Citizen science holds promise to help redress the erasure of phronesis, which has come at the hands of a preoccupation with reductive ideas of episteme or universal knowledge alone as foundational to expertise.

Defining Expertise

A theme that emerged was the capacities of experts to help people, to care about solving problems for others, and to make meaningful contributions. Chris, who identifies as a citizen scientist and a scientist, said expertise requires "commitment, caring and duty to pass on the expertise." Dylan, a scientist, says that it is important to have a "good solid knowledge base about a particular subject" as an expert, but adds that an important capacity is to "contextualize." Dylan explains that this contextualizing allows one to "get new knowledge and you can incorporate it into the knowledge that you have and shift your opinion." Importantly, this capacity, Dylan continues, means that "you're able to help people and solve problems" even when the context is new to you. Although the problem-solving here may appear to be the key capacity, helping people is the key purpose. This is to say that problem-solving is in service of some end, and many responses by interview participants remind us that it involves working with others. Blaine argues that although expertise requires some "mastery of the discipline," that "more than that would be the ability to communicate that expertise to other people and then to help them solve problems that lie beyond their own expertise." Sometimes this application in the service of others is achieved through transformation of knowledge into practice-as-solutions. Reese, a citizen scientist, explains that expertise, particularly as related to research, "refers to knowing some sort of information and making that information actionable, so being able to do something with it," a remark that echoes the practical application others have mentioned, adding, "Even if it's just explaining something to someone in a way they can understand, then that's expertise as well." Reese further connects the question of expertise and helping others, explaining that expertise involves "knowing the procedures necessary to accomplish a certain goal and having a knowledge base about the field," adding that, in addition to knowing procedures (a kind of techne, perhaps) and knowing the field's knowledge base (episteme), expertise also means "that you're able to accomplish meaningful outcomes—be they research outcomes, connecting people outcomes—and just the ability to build something meaningful in the field, based on your knowledge of it." Such judgments require, we might now say, understanding the relationship of phronesis to other forms of knowledge, other capacities for deliberation and judgment. One manner by which we can cultivate these capacities is through experiences, which build via memory and recollection into a kind of practical wisdom.

"Expertise means that you have experience," Ashley explains. Ashley understands there to be two major features of experience: "One is like it's experience and it's wisdom, but also, it's sort of technical expertise in something in a more sort of programmatic data science kind of way." The manner in which we can identify the benefits of such experience is in the level of difficultly one faces in finding a means by which to respond to a particular solution: "The thing that would take me many hours to figure out how to do would take you ten minutes." Ashley elaborates that "it's like you have a sort of lived experience to acknowledge to know what it takes to do this kind of work and also what are the obstacles that you will come up with, but then also that you have that well-read perspective; you have the sense of what is possible." Interestingly, Ashley specifically invokes the notion of wisdom, and in relation to experience. Having some sense of how experience cultivates a kind of wisdom illuminates why this combination might result in the apparatus for efficient deliberation that Ashley notes, and also the necessary capacities to help people solve problems and find meaningful solutions to said problems. Powerfully, Grey notes a key idea about this capacity of expertise, which is that being an expert is often aspirational: "To be an expert is something that you can never reach, but every time that you go outside, you are getting closer and closer."

Varieties of Expertise

Citizen science participants often noted that there are many different types of expertise, and in their answers they also provide insight into how expertise is perceived outside the strict disciplinary confines within which professionals normally operate. For instance, Drew, identifying as both a citizen scientist and also a scientist, explains that "there's so many different contexts and kinds of expertise." Drew tells us that expertise can mean "having the skill, knowledge, and ability to be able to do the task at hand or reach the goal." Consider the following example that Drew provides: "If you say, 'I'm an expert birder,' there's some expectation of [what] you're capable of. Identifying a lot of birds, knowing what birds are typically found in this habitat." Notably, this is also contextual information (knowing what is in a particular habitat), which relies on theoretical as well as practical knowledge. River, a citizen scientist, says that "expertise means that the people are very knowledgeable on a subject. They just know a lot about it," but elaborates that "there's types of expertise." River distinguishes between a "kind of general expertise" and a "very specific" expertise. For example, River says, there is a "generalist expertise and just kind of broad knowledge of their subject and how other things tie into it. And then there's the real niche experts." Such a distinction, again, invokes the idea of Willie in his shop (see Harper 1992) and more niche-based work, such as many of the self-descriptions of research areas given in the last chapter from experts working as academic research scientists.

Sean, a citizen scientist, makes another distinction between forms of expertise: "I would say expertise is probably two kinds, formal and informal," arguing that "if you have a PhD in something, society recognizes you as an expert in that field.... If you have a PhD in beetles, you're going to be considered an expert in beetles." Another kind of expert, the kind found in citizen science, describes "where you have laypersons who are trained and supervised by scientists. They form another form of informal expertise." Devon, a scientist, also provides a more expansive definition of expertise than was typically found in interviews with scientists and researchers who were not involved with citizen science. "I would say expertise means being one of the most knowledgeable persons on a topic," Devon says, continuing that, however, expertise "can be on any topic." Devon provides an example, once again recalling the distinction between expertise and specialization: "My mom can be an expert on this way of cooking paella on a Sunday. While she can also be an expert in statistics because she's a professor." Indeed, cooking is an excellent example of how we imagine expertise, not unlike Willie, involving various kinds of knowledge required to perform expertly. An important socially situated dimension of expertise, however, is emphasized with cooking, which very much is culturally defined. This, too, reminds us that expertise broadly is often socially or culturally situated, and that phronesis is contingent (like techne), not universal.

Theoretical and Practical Knowledge

Ryan, a scientist, provides a conventional understanding of expertise, saying that the term, "to me, means that you have—to put it simply, you have a PhD. I think the definition of a PhD in general is that you have spent a considerable amount of time focusing on just one thing." Reaffirming credentialed notions of expertise, Ryan explains, "I wouldn't call anybody an expert in something unless they have a PhD in something or they have an extremely dense and long amount of experience in one topic." Providing an example of someone who might have extensive experience rather than a PhD, Ryan explains that someone "might not have a PhD but they've been researching the same topic and doing things around that same topic for over four years. That's what I would say defines expert. Either a length of experience that shows considerable expertise or some kind of certification, I guess, in a way that shows expertise." Experience opens up who might be defined as an expert. Hudson, also a scientist, explains that an expert is "someone who is extremely knowledgeable in whatever area it might be," with many people obtaining this kind of knowledge through formal education. However, Hudson notes the importance of informal training and experience: "I've also been equally blown away by people when I'm outside pooling who would just go tide pooling[5] every day and know every single critter out there. So I think you can obtain expertise in a lot of different ways but it all kind of boils down to just having a lot of experience in that area." Sean offers an example of how expertise in citizen science projects is not always demarcated by formal training or credentials, explaining, "I used to dive with a . . . marine biologist" who was completing a doctoral degree in the field and had considerable training as a marine biologist:

> And at the end of every dive I would see some sort of marine life, and I had no idea what it was, and I'd ask him. I'd be like, "This is so great. I'm diving with a marine biologist. He'll be able to tell me what I'm seeing down there." Every single dive, I'd say, "Hey, Brad, what was that?" And he'd go, "No idea." And we'd do another dive and I'd go, "Brad, what was that?" And he'd go, "No idea." And this went on for three or four times and finally I said, "Brad, there's something I don't understand here. You're supposed to be a marine biologist, but you don't know—you can't identify anything on any of our dives." And then he explained to me the nature of doctoral work, which is the old joke about you study more and more about less and less, and finally, you know everything about nothing, right?

Sean continues this story, noting that Brad's doctoral work investigated a specific and difficult to find species of underwater life. "Now, contrast that with some diver that I know who have no formal degrees," Sean says. Although these divers have no formal training, Sean notes that they have spent decades studying a particular sea animal. With such experience and dedication, Sean argues, the divers'"level of expertise [with their subject] could rival or exceed that of, say, my friend the marine biologist" who had never studied these other sea animals.

Social Construction of Knowledge

In another formulation of expertise with an expansive understanding of experience, Ellis says, "Expertise means to me both a familiarity and a comfort in the topic. So someone who has expertise in something is someone who has learned about it, has studied it, but also practices it." Ellis, however, explains that "developing expertise in a topic isn't just taking a class on it." Instead, developing expertise is a matter of practice within a community. The process of becoming expert, for Ellis, is one that requires "becoming very involved" and "asking your own questions about that topic and answering those questions." In the example of a scientific subject, Ellis explains that one can conduct either their own research program or participate in someone else's. In both cases, what allows someone to become expert is "being there and present and involved in that topic." This presence and involvement, Ellis concludes, "is what I think develops expertise." Indeed, recalling Collins and Evans's (2007) account of how expertise is developed, there is an important function in immersion within a community of experts. Putting expertise down to mere acquisition of knowledge would be reductive, we know. But knowledge must also be broadly understood to mean more than episteme or even techne. Forms of knowledge necessary for expertise also include the kinds of social knowledge acquired by participating within an expert community. Participation of this kind affords both overt heuristics to assess experts as well as tacit knowledge about how to conduct research that is often gained through immersion. Indeed, citizen science, unlike other forms of engagement with science, is interesting for this reason. Becoming expert as a citizen scientist requires, in a similar manner to becoming a professional expert, socialization. The enterprise of citizen science is important in this respect for providing a community of other experts.

Defining Expertise Across Disciplines

When we asked participants if these ideas of expertise are applicable to other fields, they provided interesting insights, often noting without much concern that there is indeed variation across disciplines. Often the idea of expertise is focused on the kind of knowledge we might call *episteme*. Disciplinary expertise, the content knowledge we often imagine in the process of formal education, is central to understanding a problem. It shapes how we understand the problem. Ashley notes that expertise likely has distinct features across disciplines, fields, and specializations. Expertise is not a standard notion, Ashley argues, as it is "different because different disciplines have different standards of expertise." By way of example, Ashley explains that the very idea of collaboration reveals the differences in understandings of expertise: "The musicology field will consider expertise but that won't mean anything to the biologist. But there are projects that pair ethnomusicologists [and] symbiologists when we're talking about bird song or we're talking about music and sound in 'nature.'" In such collaborations, much of the work Ashley has undertaken has been "finding those commonalities, those common grounds" rather than focusing on each specialization. Notably, here, the relationship between episteme and the process of socialization into a discipline are so closely interlinked that one must understand both to understand how episteme is constituted as knowledge; that is, socialization is a key to understanding disciplinary epistemology. Collaborations often point to broader distinctions among forms of expertise across the disciplines. Blaine, a scientist, argues this point, saying that conceptions of expertise are varied, noting that their colleagues "in the communication disciplines have a somewhat distinct set of skills and ways of exercising them."

It is not only in different disciplines or collaborations that such distinctions are found, but even within fields, where there are both theoretical and applied research trajectories. "It definitely means different things across different types of research," Ryan, a scientist, explains. Notably this is so because "some types of research in academia are very applied as in ... we already know everything about it and it's just a matter of learning the things that we know. Some things like biology are full of question marks still." Within disciplines or areas of expertise, there are also distinctions between formal expertise and local or traditional knowledge. Recall, for example, Wynne's (1989, 1992) sheep farmers. They had some knowledge of the nuclear industry and how to interact with it, how to trust or not trust their representatives through forms of social knowledge (see Collins and Evans 2007), as well as the local knowledge of their farming practices. These important distinctions remind us that conceptions of expertise must account for more than formal training and credentials, or sanctioned forms of knowledge, as well as the critical socialization into a discipline.

Devon, a scientist, believes there is a difference in how expertise is understood across disciplines: "I would say in the natural sciences field, the people with whom I'm working, botanists specifically, consider expertise as knowing, for instance, the botanical name of a plant. . . . While probably there's other disciplines, for instance, more like ethnoecology or some other more transdisciplinary fields, where expertise means other things." By way of example, Devon explains that "in ethnobotany, you can have a local expert. That doesn't mean it has a botanical or any kind of formal education." Another response, from Rory, a citizen scientist and scientist, notes this distinction between forms of expertise based on knowledge, saying that they do not believe that expertise is a standard notion across disciplines. Rory believes that "different types of expertise" can include those with significant specialization, giving us the example of "butterfly experts [who] don't know a wasp from an ant." Hudson, a scientist, explains that the question of whether different fields have a different understanding of expertise is "kind of tricky." Overall Hudson believes that if one identifies or is identified as an expert there will be certain expectations about what they will *know* but concedes this is difficult to precisely describe because there will be some variation across domains. The relative age of a research field might also shape how much and what form of knowledge one is expected to have as an expert. "Different areas of science," Hudson explains, "have a lot more previous research behind them so, in order to be an expert in a field that is really advanced, you probably need a lot more background information than an emerging field." For multidisciplinary teams, this perspective is both interesting and challenging. Knowledge, here, seems to be strictly used in a conventional sense of epistemic knowledge. Yet, working in an emerging field, often interdisciplinary in nature, requires different forms of knowledge and capacities to integrate those knowledges.

Different forms of knowledge, however, may be the common threads across disciplines. Ellis believes that epistemic forms of knowledge may be distinct across fields, but that "more broadly, the definition of [expertise] is relatively the same." Ellis asks us to consider the example of becoming a legal expert and an expert in dance. For Ellis, "the definition of expertise across the many different topics stays the same in terms of both *familiarity and comfort and experience* in the topic." Ellis's point is one that Ericsson and Pool (2016) themselves might have written: "Experts are not born overnight by any means. I think it's a very *time-consuming* thing to become an expert in something." Max, a citizen scientist, also notes the temporal commitment required of becoming expert: "I don't think I have expertise because I do it as a hobby. I do it casually, and I don't wish it to be a job or to have the pressure, to stay on the cutting edge with what's going on so that I could explain to people what it is I'm involved in." These comments begin to move the definitional questions of expertise into a space to talk about assessment heuristics. Yet these heuristics may not necessarily operate across all domains of expertise.

Sean explains that different communities of expertise will define and value expertise differently, beginning to move toward questions of assessment of expertise. When speaking with someone who works in an academic setting, this person might tell us that you need a graduate degree in the field. "However," Sean argues, "if you talk to citizen scientists, we will say what I've been saying, which is that there's another kind of expertise, which is informal." By "informal" Sean means cases "where you don't necessarily have a degree, but you have proper training and supervision by a scientist." Understanding expertise in this way is in line with the kind of model that Collins and Evans (2007) have established. Participation in the community of scientists, in the field of practice, and thus expertise can be cultivated in a manner that allows for interactional and contributory expertise among professional scientists and citizen scientists. Citizen science is a well-established enterprise with a rich history, but how practitioners trace that history depends on how they define who is a citizen scientist: "We like to say Charles Darwin was the first citizen scientist," Sean tells us. "But let's say in more recent times," Sean continues, that "citizen science has really only been a topic of discussion for about a decade." Over the last ten to fifteen years, citizen science has certainly taken on new forms and generated considerable interest—and skepticism. Sean tells us that although there was initial skepticism, the work citizen scientists have been completing and, crucially, publishing about in peer-reviewed journals has helped to change that skeptical perspective. Again, Collins and Evans's (2007) model is instructive here as it underscores the importance of a community of experts developing their own meta-criteria for assessing expertise. Collins and Evans's model also helps us understand the perspective of Dylan, a scientist, who tells us about expertise across disciplines. "Different across different disciplines?" Dylan repeats back to the interviewer, reflecting on the question about how expertise varies, before adding, "I think it can be—it's that debate." Dylan elaborates, first asking, "Is it how much stuff can you do? Jack of all trades." Then Dylan reframing the key question: "You can be good at lots of different things, but can you really excel at all things?" From Dylan's vantage, "the answer is really, strictly, probably not." However, just as we want to expand the kinds of knowledge that constitute expertise, it is important to expand the kinds of expertise we look for among experts. Although the answer is "probably not," Dylan argues that this "doesn't mean that you can't have an expertise that involves integration." The kind of expertise here might be a sort of interactional or referred expertise.

Assessing Experts and Expertise

Expert Status and Trust

Trust is among those tools used to assess an expert's credibility. There is a certain vulnerability to this idea, as Reese, a citizen scientist and specialist in communication, notes: "I guess they're an expert until proven otherwise, right? So you have to trust people's word for it until they do something to violate that trust." We also asked Reese about what the public might not understand about expertise. This is not a response detailed in this analysis, but Reese provided an important insight to our questions of good or moral comportment: "I think people feel very threatened by experts. So I think maybe what people don't understand is on an individual level, they may really like an expert and they may really appreciate that expertise." Reese adds that "in the abstract sense, I think people think experts are cold or they're looking down on them, having a certain condescension. So I think people don't understand that a lot of experts become experts because they want to *do good things* with their expertise," adding, "I mean, no one thinks that I'm going to become an expert so I can condescend people all day."

Trust is gained through several mechanisms, including credentials, as Max, a citizen scientist, explains. Max sees several factors in addition to credentials, however, as being important: "I would think that I would have to have read information that they have shared about it, or I would have watched an interview or seen some type of report from this person, that would appear, that they know what they're talking about." Max warns that "you have to be careful," and that credentials may be "something that would make a person feel confident that you are getting information from someone who has expertise or is an expert." Trust, or earning trust, is based on one's audience's assessment. River, a citizen scientist, assesses a prospective expert's credibility by listening "to what they're saying, and I listen to see if it makes sense to me, and I really pay attention when somebody asks [them] a question to see what their answer is." Adding a strategy within this approach that helps to assess goodwill is to tell the expert about something you saw and say, "If they dismiss me, your credibility's gone." Part of good research, River argues, "is when you're out researching an area for something, talk[ing] to local people." River's comments here demonstrate the crucial role of goodwill in citizen science projects, explaining that "I might not know what a [inaudible] is, but you mean those little owls with the long legs? Local people pay attention, even if they don't know what they're looking at." Cases where goodwill is not shown cause River to judge the credibility of a purported expert more harshly. River explains by way of example, saying, "I've come across guys who are birders and . . . they'll dismiss you when you say something," qualifying that "they might still be experts" but because of their lack of goodwill suggests an undermining of expertise: "these guys can't accept that they're wrong, or if somebody else has a better idea, or another observation that if they haven't written this up themselves." Further to being more covertly dismissive, there are also those who are overtly dismissive of others' experience, which, predictably for those with a modicum of rhetorical sensitivity, diminishes credibility. River explains that "I've actually worked with people that said, 'Hey, I've got a degree. You're nothing but a dumb technician,' [that] type of stuff. So I guess those people were dismissive of what your life experience might be, or your own personal observation."[6] What such dismissive postures produce, other than doubt in rhetors' goodwill toward their audience, is, in fact, a problematic stance toward knowledge production. Their affective condition, perhaps insecurity or vaingloriousness, causes them not only to lose the goodwill of an audience but also to overlook perhaps important insights or information, due to excessive attention to attributional qualities of expertise grounded in credentialing rather than experience-based accounting of expertise.

Drew, a scientist and a citizen scientist, argues that "there's little clues when you're talking to someone that their ideas are consistent or make sense or resonate with stuff that you've heard before" as one means to begin assessing experts. "There is the degree [of] like formal, legitimacy, trusting," Drew continues. "There's the trust—or that they're [a prospective expert] already—maybe they're like a leader in an organization that you already believe in the organization. So *they're kind of borrowing their legitimacy from an organization.*" Drawing from the field of birding, Drew provides an example: "Maybe they're the director of education for Audubon or something and you're like, 'Yeah, I trust Audubon. So I trust that they would hire a good director.'" Institutions are not the only system for legitimation; as professionals also noted, one's professional and social networks are crucial to the assessment of expertise. So, too, is the case for Drew: "And then, there's the network thing that someone that I know trusts them. And so you're willing to extend that you believe that they're an expert." Drew demonstrates several tracks for determining expert status and expertise, including trying to assess individuals' knowledge relative to your own, their credentials and track record, as well as their situation within an expert network or organization that provides credentialing. Among Drew's comments, too, are those that signal values, including *trust* that one is an expert, *trust* in an organization, and *trust* such an organization would hire a *good* director. Critical capacities, as discussed throughout this book, are often born of not only technical knowledge but also experience, including with the kinds of edge cases that allow one to develop the kind of discernment to judge how best to respond to a situation. Understanding and being critical of theories and best practices is a good example of knowledge extending into this realm of expert discernment. Grey provides further insights into this line of thinking, explaining that when beginning to assess a prospective expert it is important to understand that expertise requires not only content knowledge but also the capacity to learn and grow as an expert, saying that in the case of identifying different species, the "knowledge of the identification of the species is not mandatory but the willingness to learn, to participate, to collaborate with other people, to be available to understand your own limitations" is important. Understanding the limits of one's knowledge has been noted in a variety of expert models as essential. Combining an understanding of one's own limited knowledge and the capacity, and interest, to learn is essential to becoming expert. Grey, however, also reminds us of the importance of community in the acquisition of expertise, not only in the kind of socialization model, but also in terms of an ongoing negotiation of knowledge. For example, Grey elaborates, identifying species and subspecies can be quite challenging, even when one has considerable experience and expertise. Collaboration among experts allows them to not only pool their relative knowledge of classifications and skills in identification, but also demonstrate to those with less expertise or experience on a team the processes by which their expertise proceeds. A recurrent theme in the identification of experts was that they had some capacity to share their expertise.

Expertise as Instruction

Bailey, a citizen scientist and scientist, explains that in the assessment of a prospective expert, one ought to determine the candidate's "ability to explain what they know in common, plain language." Understanding subject matter at the level of an expert, this line of thinking goes, is rich, deep, but also so complete that the analogical work required to explain the subject to a nonspecialist can be reasonably well accomplished. "If they can't tell a story that their grandmother or mother would be able to follow," Bailey says, "they don't really know it." Bailey's comments are not judgmental, however, but rather reflective: "I know I'm in the same boat. If I don't feel that I can explain something clearly enough to somebody who's ten years old. . . . I look at myself and say, 'Okay, maybe I don't know that particular species or that particular concept well enough.'" In addition to demonstrating one has mastered content or skills, however, there is also a kind of expertise that needs to be developed to successfully accomplish this kind of analogical work or otherwise accommodated discussion of a technical subject. For Bailey, it took time to become expert in their area and also in this other skill of communicating technical subjects to nonexpert audiences: "I would admit that probably it took a while for me to get out of the jargonistic approach that even while I was doing consulting law. . . . A lot of that information was inaccessible to the knowledge of the public because there was too much jargon." With some effort, Bailey notes, "I tried to beat that out of myself and work with other people to make sure that language is accessible to a broader audience." Fahnestock's (1998) notion of the accommodation of science reminds us that simply removing complex terms does not make technical subjects accessible; rather, it is the attunement to the audience and their purpose for the text that makes them so. Bailey's own motivations are in line with such efforts: "I would find a way to make sure that the language was accessible to your average reader because otherwise, the reports are useless." Phronesis is implicated here through the *ethoic* work accomplished in an accommodation of technical knowledge with the function or purpose of that knowledge for the reader (average, broad, or otherwise) in mind.

Eddie first cites looking at someone's professional status or credentials, and then "you kind of apply the sense test." By sense test, Eddie means that someone appears "to understand the material," but cautions that "this is a little bit misleading on occasions because not everybody's a great communicator who's an expert." However, Eddie adds, "In general experts can effectively communicate the basic principles of areas in which they are experts," so using a sense test can be a helpful strategy in assessing someone's expertise. Euroa, a citizen scientist, provides some more insight into the ways that one can discern whether purported experts seem to be able to make a sensible case for their areas of interest. "Usually, they have data to back up what they're saying," Euroa notes. "They'll either know the person who found that item or something or they will go, 'Well I've heard of it from here and it's corroborated by there and then we put a paper through to *Nature*.' Or something like that." Euroa provides an example, slightly adapted here in order to maintain anonymity. Euroa explains that there are individuals who can do this kind of work. There are also individuals who have impressive local or traditional knowledges that can greatly improve scientific understanding. In a citizen science project, a group of individuals has shown that a sea animal once believed by the scientific community not to travel far from home in fact does. The story follows that a young biologist learned at university that this sea animal does not travel far from its home, but her father had observed one of these animals on distant sides of the island on which they live. "And they went, 'That's weird, but how do we prove it?'" Euroa tells us. After speaking with scientists about how to do just that, the citizen scientists were told they would need professional help to complete the study. Despite this caution, the citizen scientists went out to complete the work and got tourist operators involved, and people took many pictures to identify this animal and its travel. In the end, the data were quite good and have resulted in nearly two dozen publications, some of which appeared in the very top science journals. The group now has something of its own affiliation, and publication record, so even in citizen science sometimes these heuristics prove a useful measure of expertise. Importantly, in such cases, we might note that the expert status achieved, and the expertise required to do so, were hard fought, not owing to established models of credentialing. Although there are attributional features of expert status and even expertise, citizen scientists reveal that even when those attributions may not initially be made there are still processes at work to become expert.

Research Record and Affiliation

Perhaps not surprisingly, one's track record and affiliation played an influential role in the assessment of expertise. In citizen science, however, a more expansive understanding of track record than one's grant and publication history is necessary. Hudson, a scientist, explains that "in the academic setting I definitely trust the peer review, and reading journals, and things. Knowing that it's published in a reputable journal can really say a lot." In the case of citizen science, such measures are not always available or appropriate. Hudson has strategies for assessing expertise that remain skeptical of uncredentialed prospective experts but still allow for an acknowledgment of other kinds of expertise: "In just meeting somebody more casually, if they identify as an expert, I'm definitely going to try and be a little skeptical and kind of try and figure out where they got their experience from before I myself consider them an expert." Blaine, another scientist, relies on what we might think of as more conventional measures of an expert but acknowledges that such assessment "depends on the circumstances." In cases where Blaine does not know the individual well, "a lot on credentials" and "how they're regarded by their own peers" would be important. Credentials would include "grants, publications, their academic careers, what sort of a school they work for," adding that in the latter case, the status of the institution would be important. Another strategy of Blaine's to assess expertise is to read publications by the prospective expert, and Blaine explains that you can tell a lot by reading these papers. Blaine adds that, if these strategies fail, calling other individuals in one's network—

"people I know and trust"—is a useful approach, but not infallible. "I get fooled sometimes. I've been fooled," Blaine adds.

Parker, another scientist, reveals the challenge here, telling us, when we ask how they assess an expert's status or expertise, "They know more than I do [laughter]." Joking aside, Parker explains that "somebody who's probably been working in the particular field for probably five or more years [and] has peer-reviewed publications or books" is likely a good candidate for having some expertise. Or, as Parker explains, such credentials certainly help "build the credibility to make you an expert in a subject." Ashley, a citizen scientist, uses an established social network of friends as well as publications to assess credibility. A colleague of Ashley's reached out to a work group Ashley is involved with and explained that their friend was offered a job by someone. Was this someone "legit," Ashley's colleague wanted to know. Ashley and the group examined the website of this individual, noting "he had a bunch of credentials, but it all seemed sort of like he made them up." Credentials were undermined by the perception of a large degree of self-promotion, which may be a warning sign: "The fine line that experts need to walk between promoting themselves—promoting themselves, especially if they're in an academic setting, for sort of tenure and impact and things like that—and then also sort of the con artists out there." Even in those cases where there are publications, institutional affiliations, and so on, "those things aren't foolproof," Ashley explains, and thus situating this information through one's broader professional and social networks becomes important: "it's going to colleagues and peers" to get a "behind the scenes" and look into a situation. For example, Ashley suggests, "this person has an appointment here, or they're not really well respected within that department or within that field," and this "triangulating" of different information provided by credentials, publications, and professional and social network insight is crucial in the assessment of expert status and expertise.

What Citizen Scientist Experts Reveal

When Hartelius (2011, 6) asks, "When expertise is an attributed state, on what does the public base its attribution? . . . Are citizens' attributions of expertise valid? Can the public make discriminating judgments distinguishing experts from imposters?" the questions are posed to a citizenry without the benefit of understanding expert paradigms. In science, we can see how these questions might be asked anew when considering citizen science. Citizen scientists provide important insights into expertise and expert status because they operate outside, at the boundaries of, and inside scientific spheres. I argue that this is because citizen science is, in many articulations, demonstrating the importance of the kind of rhetorical, moral knowledge discussed in terms of phronesis, ethos and its constituent parts, and the broader ethical comportment toward one's audience. Here the goals of democratic, connecting, and ethical relationships are what citizen science can offer and what many of our participants suggest when they speak of trust or even the ability of an expert to be able to teach or explain their position.

When speaking with professionals, the importance of knowledge as episteme and experience allowing for skill was evident in many responses. Hudson, recall, told us that an expert is "someone who is extremely knowledgeable in whatever area it might be." Many participants, however, allows for variety in the kinds of expertise that are possible. Drew commented, "There's so many different contexts and kinds of expertise" and Ellis, further complicating ideas of expertise, noted, "Developing expertise in a topic isn't just taking a class on it. . . . It's becoming very involved in it in yourself, asking your own questions about that topic and answering those questions." Expertise, even in professional and academic settings, comprises the features that are here described by citizen scientists, from the necessary field-specific knowledge, the varieties of expertise that might exist within and across said fields, and the kind of social immersion required to fully understand the content knowledge and skills of a field as contextualized through its social, discursive, and moral norms. Those features that I have argued are central to experts, including phronesis and eunoia, are emphasized by some participants, too. "It's experience and it's wisdom," Ashley previously told us. It is perhaps not surprising that those involved with citizen science projects would note the relational aspects of expertise. Chris, for example, noted the importance of "caring" and "commitment" and Dylan said experts should "help people," while Blaine explained that experts work with people to "help them solve problems." Reese noted that helping to solve problems can be quite simple, too, saying that someone can demonstrate their expertise in this way "even if it's just explaining something to someone."

From all our participants, several lessons that likely sound familiar have been identified in the process of becoming expert. Learners should not be passive, and this means they must take ownership over their learning. Learners, however, must be helped to develop the kind of critical thinking capacities that will allow them to fully embrace this idea of being "active" and engaging in lifelong learning to continually be expert. Avery's comments about how in mathematics there is a phenomenon where once one sees the solution to a problem it makes sense but that does not mean the same person could solve the problem, is an interesting example of the challenge of "critical thinking" and also broader understanding of problem-solving. Avery tells us that this suggests, to them, that to become expert involves the kind of critical thinking that will allow someone not only to understand why something makes sense, but also to explore why something does not or did not make sense. By being able to examine where procedural knowledge failed in a process, they will be able to engage in the kind of critical reflection that allows for more adaptable thinking born of experience. Related, learners must be able to integrate knowledge, and this kind of adaptable thinking facilitates such integration. When speaking of integrating knowledge, this might be of multiple kinds, including theoretical, practical, and the forms that emerge from identifying limitations, failures, as well as edge cases. While learners must, according to many of our participants, master the literature of their field, this mastery emerges from a critical engagement with the literature and its attendant arguments and the charting of a field or specialization's historical and future trajectory. Undergraduate and early graduate level students benefit from course work and assigned readings. However, more intensive engagement through comprehensive exams and, later, publishing marks the kind of process whereby one learns not only the literature in their area but also begins to demarcate the limits of said area. As Quinn reminded us, academic disciplines (and we might add other professions) are not "coextensive with expertise." However, when we look for ways to identify expertise, in academia, a "marker of expertise is the ability to publish in the academic journal," Quinn noted. Academic journal publishing is an interesting avenue to think about the process of becoming expert because it allows for another feature that several of our participants noted as important, and that is failure and, in this case, relatedly pushing one's limitations. While there are many regrettable features of academic publishing that one might critique, it also offers important paths for researchers to become expert. The ability to publish as a mark of expertise requires, minimally, that individuals have good mastery over the literature of their field such that they are able to identify a problem within a given territory[7] and situate their own work within the trajectory of the disciplinary conversation. Although this kind of publishing may seem to demonstrate proficiency in theoretical matters, there is a highly practical or applied element to this work, too, insofar as the transformation of research into a practical contribution to a disciplinary conversation is a significant rhetorical matter. Here the learner not only masters the technical or theoretical matters of some area of specialization, but also demonstrates a kind of expertise that many citizen science participants noted. Participants discussed this kind of expertise in terms of explaining concepts, problems, or ideas to someone else (admittedly in this case a more technical audience than they perhaps meant, but an audience nonetheless). Failure is ripe in this model, as are the kinds of learning experiences to push one's capacities born of this failure.[8]

The idea of expertise in a rhetorical framework necessarily accounts for the relational features of expertise and expert status, but, crucially, rhetoric reminds us of the social action such relational performances underlie. No longer distant, in their lab or otherwise removed from the spaces most find familiar, *citizen* scientists are tethered to the *polis*, broadly conceived, and inflected throughout with a sense of civic comportment. Citizen science is a compelling case, as I have suggested, to unravel the seemingly clean expert/nonexpert division. It reminds us that professional is not synonymous with expert, and that for all the formalities and credentialing there are other ways to master subjects, to bring together knowledge and experience, and indeed to do so with the kind of practical knowledge and moral duty that phronesis recalls. Importantly, adding notions of informal and local knowledges and questions of how credibility and expert status are negotiated outside the world of credentials, institutional backing, and publication records unfold conceptions of expertise that complicate easy binaries of expert and nonexpert, of knowing-that, knowing-how, and knowing-why.

Conclusion | Cultivating Expertise

Becoming expert involves the process whereby we build knowledge and capacities and then perform as experts in those social roles where, as experts, one is called on to assess and respond to some situation with specialized knowledge, capacities, or skills. To do so we must cultivate certain capacities within ourselves and embed ourselves in expert communities. A virtue ethics approach provides a critical perspective on how moral knowledge is co-constitutive of how we act in the world even in the generation of scientific knowledge and other technical knowledges that, especially beginning to take hold during the Enlightenment, were seemingly absolved of the moral comportment our current moment demands. Restoring phronesis to the conversation concerning the forms of expert knowledge helps reorient expert duties to problem-solving beyond mere technical proficiency, inclusive of episteme and techne. Instead, we see that practical wisdom is critical to the relational features of expert knowing and the operations of expert capacity. Plainly, experts acquire their status not merely by knowing-that (episteme) or knowing-how (techne) but, crucially, through their relation to nonexperts and enactment of prudence through an appropriate and well-reasoned deliberation, judgment, and action in response (knowing-why). Expert knowledge is important not simply based on what the expert knows or what kind of tasks the expert can perform, but also for whom they relate their knowledge or for whom they perform their tasks. Phronesis, with its morally inflected understanding, is crucial to understanding this relational aspect of the expert's knowledge.

Without such an understanding, the lack of trust in experts seems perplexing, a problem of expert and nonexpert social dynamics that somehow pull an expert from the technical sphere into the public sphere (Goodnight 1982). Distinguishing the expert's technical knowledge as episteme and techne in a technical sense from phronesis (and trust) in a public sense is only rhetorically sound if we take phronesis to occur only in public spheres. Phronesis is indeed crucial to expert-to-nonexpert engagements, but also among expert-to-expert engagements. Experts interviewed in the course of this study recognized the importance of phronesis among expert discourse. So, too, has a long tradition of social studies of science and social studies of experts in the sciences, specifically, recognized similar concepts as critical to expert-to-expert discourse. Collins and Evans (2007) give much attention to the discursive or, here, we might say rhetorical norms that an interactional expert must obtain to gain expert status. However, these rhetorical norms include forms of deliberation and reasoning as well as more superficial notions of "talking the talk." A rhetorical reading of these interactional experts shows a rhetorical savvy that is not merely a comprehension of a technical vocabulary, for instance, but a broader understanding of argument norms within a particular discourse community. Arguments are partially constituted by the very situations of which we conceive, and how we conceive of a situation is shaped both by the rhetorical worlds within which we are embedded and by our phronetic capacities. So, too, are our responses. In addition to the technical vocabulary of some scientific discipline or another, there are pervasive narratives within science—"brilliant jerks" and the like being particularly harmful narratives—that typify the responses one might have. A new conceptualization of expertise, one that centers moral knowledge as a practical knowledge for scientists or other researchers, for technical experts in industry, contends with those conceptions that excuse or even exalt bad behavior.

Phronesis or practical wisdom is an important rhetorical concept because it reminds us that our appeals, even in some of the most technical discourses, are appeals to other people. In some cases, the appeals may be to people who are other experts wishing to advance knowledge, implement what we know, and so forth. In other cases, appeals may be to people who may to benefit from expert knowledge in the application of said knowledge to their nonexpert everyday lives. It is perhaps clearer why attention to phronesis is important when it is subverted. Sometimes it may be that scientists have a careless orientation toward their commitment to others, for example. There are also, however, those who have more malicious intent in the selective deployment of appeals to goodwill, which is, rather, a deceptive comportment. In these cases, experts lose credibility not only for themselves but broadly for their specialization or field, as they have broken the relational agreement where phronesis is expected (although rarely by name) to partially govern the application of their knowledge in ethical ways.

Sometimes, however, goodwill, in the sincerest appeals, can undermine an expert. Experts, in many configurations of the term, help manage uncertainty for nonexperts. We, as nonexperts, look to the scientist, as Pietrucci and Ceccarelli (2019) showed, to help us understand the risks of an earthquake. Even the common forms of popularization are rich with appeals to help us ease uncertainty. How much coffee can I drink? How much money should I put away for retirement? Turning to experts and expert knowledge to answer these questions is reasonable. Generally, experts can provide reasonably well-argued rationales for, say, drinking coffee or planning for retirement. But sometimes the uncertainties we face are uncertainties for experts as well. Consider, for instance, end of life. A doctor may tell someone they have an incurable disease and offer them the option to try to prolong life, but it will mean months of devastating treatments that, ultimately, will make the patient even sicker. Another option, the doctor may say, is to refuse treatment, but the patient will then have only days to live. Several individual, familial, and social decisions must then be made, but even once those decisions are made, we still turn to medical professionals to help manage uncertainty. If the patient, for instance, refuses treatment, how many days do they have and how is end of life, then, managed? A doctor may be tempted to provide a range of days, for example, that someone has left to live if the patient is presumed to be near death without any treatment. But this is not a certainty, and the complexity of each patient normally means they will pass in what is a rather unscientific measure: their time. Although a doctor may be trying to provide a family with some certainty, some ability to plan and manage end of life, such a goodwill gesture can undermine confidence in the doctor as a medical expert. This, however, is in some ways a question of understanding the limits of one's expertise. Except for palliative care medical professionals, physicians are often experts in healing, or at least trying to heal, people. Palliative cases require a different kind of expertise, one that helps manage rather than dissolve uncertainty, which calls for a different kind of knowledge. Knowledge, here, is not meant in an abstract sense, but quite pragmatically. Many of the resources physicians have to manage uncertainty in patients receiving active care relies on medical tests; however, in the case of patients with no active care due to a palliative plan, this may not be the case. Although the desire to alleviate uncertainty is, to some degree, charitable and certainly an effort at goodwill toward one's audience (the patient and family in our example), it is not difficult to understand why such an affective rhetorical stance toward audiences may fail. In cases where only a few days are given but the patient continues to live on for weeks, perhaps months, the emotional regulation (in a nontechnical sense) of the patient and family may have been initially imbalanced. With the constrained temporal horizon of a few days, goodbyes, stories, legal matters, and, crucially, experiences are compressed. A persistent and intensive anticipatory state of dread looms, too. What begins as goodwill, without a kind of phronetic sensibility for the expert—one that accounts for the situation, including the limits of a particular expert to speak to all the aspects of the situation that demand a response—ultimately leads to a failure to enact goodwill. Failure here results not from poor intentions, but from a lack of phronesis. Outcomes are not the only measure of the expert, to be certain, but here the outcomes did not fail because failure is a necessary possibility, but because of the expert's inability to appropriately measure a proper response to the situation.

Centering phronesis in the theorizing of expertise serves to demonstrate two important capacities of expert thinking. The first is a kind of practical knowledge to make good decisions. The second, intrinsic of the first, is a practical moral knowledge, as a capacity to make the best decision. Expertise is born of a social relationship between someone who has certain capacities that others do not, can perform certain tasks that others cannot, or has knowledge that others do not. Because of this relationship, experts are imbued with a kind of status that affords them both power and privilege. Crucially, as well, this privilege brings with it a moral duty to those with whom they have entered this social relationship. In these cases, experts hold knowledge or have skills that govern the lives of many people, and rarely are these experts' knowledge so far removed from society that their work can be said to be outside of this relationship.

Even those conducting basic scientific research, perhaps not anticipating the relevance of their findings for a century or more, are participating in work that shapes the direction of our knowledge, and thus how we, or our grandchildren's children, will come to live. Expertise imbued with phronesis is crucial to a rhetorical conception of expertise because it centers deliberation with respect to expert decision-making in at least two configurations. First, deliberation is understood with respect to a particular situation, both in terms of revealing and understanding a situation as well as responding to a situation. Indeed, the focus on identifying a situation, its affordances and constraints in a Bitzerian sense (Bitzer 1986) but also in terms of the *kairotic* configuration of situations that Miller (1992) offers, reminds us of the complexities in this initial moment prior to deliberation on action. Second, deliberation on the most appropriate course of response or action is principled again on an expert's phronetic capacities. This is so because the expert must not simply find a situation and decide on the most efficient or technically savvy response. Rather, an expert must understand a situation and deliberate on a response within the complex moment constituted by prior knowledge, experience, and other individual knowledges along with the historical, social, and particular constraints of said moment. Further, because expertise is relational and enacted when a response is required of a situation, the expert never acts in isolation.[1] This perhaps seems a rudimentary statement. However, even those whom we can identify as successful in achieving and performing expert status might fall into the reductive thinking. For example, one might equate the scientist with science, the thinker with data, or the expert with only universal knowledge or episteme. When this occurs (and such reductive thinking is surprisingly pervasive), the expert is reconfigured into a mere vessel for data, facts, and the like, divorced from a situation and context. This is farcical but has quite serious implications to expert status.

Expert status, as Nichols (2017) argued, has come under scrutiny. Although proclaiming the death of expertise might be rather premature, certainly it would not be unfair to diagnose it as greatly afflicted. Although trust in scientific experts remains reasonably high, the partisan political polarization in assessment of scientists is well documented.[2] Globally, anti-vaccination movements target experts and expertise using a range of strategies, from seeding doubt in effectiveness to biologically impossible conspiracy theories. Although the COVID-19 pandemic has renewed the urgency of countering vaccine misinformation, the overall challenges to vaccines are not new. The pandemic has also underscored the complexity of debates about expertise and experts, with many experts including doctors and public health specialists called heroes and celebrated while these same experts have been, by others, characterized as a villains or people with little phronesis. Experts in these domains have noted how difficult if not dangerous the politicization of the pandemic, notably in mask mandates and vaccine passports. This is not the first time that experts and scientists have noted how partisan politics shape their work. In the United States, scientists argued that their research has been squashed by politicians and bureaucrats with partisan commitments, a similar refrain shared by scientists under Prime Minister Stephen Harper's government in Canada. Indeed, scientists were so compelled in the United States that they even took to the streets to protest in the March for Science and began efforts such as 314 Action to move scientists to become elected officials.[3] While some might put such partisan political divisions down to public perceptions that are quite removed from the actual business of science—of expert work—there are also cases where experts have behaved badly, thereby losing public trust. As Ashley, a citizen scientist, noted in their interview with us, "There are plenty of people who have done terrible things in the name of science,[4] and not just even thinking about Tuskegee and Henrietta Lacks . . . [but also] Flint."[5] Such culpability is not limited to the sciences, as the 2008 financial collapse in the United States demonstrates. Ongoing questions about the role of the high-tech sector in shaping our lives, from autonomous vehicles to artificial intelligence, too, raise alarms. In combination, events have caused scientists and other experts to lose the trust of their publics. Complicating matters further are misinformation campaigns designed to unfairly undermine the credibility of experts. A constellation of challenges has formed and threatens to undermine the very idea of experts and expertise.

In order to properly redress the turn away from experts as crucial participants of our deliberative democracy, it is not only facts, specialist knowledge, nor credentials that are likely to provide a way forward. Nor is it codes of ethics or championing of ethical technology or tech ethics in its variety of flavors per se. Consider the case of Joi Ito's work in the MIT Media Lab, which came under scrutiny when funding ties to the convicted sex offender Jeffery Epstein were revealed and, subsequently, Ito's status as an expert in "ethical AI" was questioned. A former graduate student in Ito's AI ethics group, Rodrigo Ochigame (2019, para. 5), describes the ethical AI movement as "aligned strategically with a Silicon Valley effort seeking to avoid legally enforceable restrictions of controversial technologies." Ochigame explains that research from the lab was ignored and the opposite recommendations from what research might suggest were advanced by affiliated groups: "A key group behind this effort, with the lab as a member, made policy recommendations in California that contradicted the conclusions of research I conducted with several lab colleagues, research that led us to oppose the use of computer algorithms in deciding whether to jail people pending trial. Ito himself would eventually complain, in private meetings with financial and tech executives, that the group's recommendations amounted to 'whitewashing' a thorny ethical issue."

Indeed, various "codes" or "commitments" to responsible or ethical technology exist, but the behaviors related to minimizing regulation that Ochigame detail help reveal why such efforts are insufficient. The matter of regulatory importance and rejecting corporate control of the commons is a matter unto itself. However, the existence of lobbying in support of that model and, more centrally, "ethical" movements that either by design or accident support a lack of regulation demonstrates their vulnerability and, thus, inadequacy. There are larger systemic and situational reasons that experts without such moral compasses seem to thrive, but this does not negate the need for the cultivation of character; rather, it illustrates the need for it. Overreliance on semi-systematic and professional solutions, such as tech-sector codes of ethics, has failed, in part, because it is not grounded in one's character and moral comportment. Aristotle believed in a unity of the virtues, which, while not essential to commit to here, may be illuminating as to why the compartmentalizing and qualifying of "tech" ethics is a phenomenon of which one can reasonably be skeptical. Investigating "ethics" in the service of undermining regulatory efforts, it is evident that the norms surrounding experts in these fields do not adequately account for goodwill, good reason, prudence, or other virtues. Not only have these fields, inclusive of their "tech ethics," failed, they have killed some, impoverished others, and undermined democratic systems. A virtuous character—habituated, knowledgeable about what is good and how to be good, and a consistent, conscious decision-making process to do good for its own sake—is a necessary remedy among experts because their role as individuals is so powerful. However, strong regulatory frameworks are also critical, as an individual may not be well positioned to intervene. A regulatory approach alone, however, may be ineffective. Consider, for example, cases where no laws are broken but deeply unethical practices proceed unchecked.

Phronesis and attention to one's character in a virtue ethics model is not merely a proposed solution here: it is diagnostic. The current problems faced in tech, or in science or health and medicine, are not merely born of new science and technology. Indeed, as we have seen, the very historical circumstances that led us to cleave off technical knowledge and skill from phronesis, in the positive formulation of technoscientific rationality, was by design. Here the epistemological basis of science itself is not under investigation. Rather, what is explored is the role of the expert—indeed such a role is a social agreement and rhetorical in how it is negotiated—and the socio-cognitive capacities (to which we ascribe the shorthand of "expertise") necessary to fulfill that role. Refiguring the practical knowledge of phronesis as a central requirement of expert knowledge is critical to the undertaking of reconstituting expert and nonexpert relations. Phronesis also helps us become better experts. Expertise is not simply something to be attained and then performed, but rather something that requires ongoing effort to acquire, maintain, and, critically, enact relationally.

Ultimately, it might seem a strange proposition to some to suggest that one is completely expert in technical domains when in possession of phronesis. But to be an expert is to exist within a rhetorically negotiated role, not merely to have specialist knowledge or skill. Indeed, phronesis is often inflected with a moral valance that might not appear a central capacity that is required to know and to act expertly. In this book, I have argued why phronesis is indeed an important capacity to both develop expertise and to act expertly. In the case of acquiring expertise, phronesis affords the ability to deliberate on the correct and good course of action. St. Thomas Aquinas provides the basis for understanding how phronesis, as prudence, is cultivated through a process of making meaningful experience, tied up in capacities for memory and deliberation. Hursthouse (2006, 285) details the enactment of a person with phronesis, saying such an individual is someone who "gets things right in action in what we would call 'the moral sphere.'" Technical or scientific spheres are not distinct from moral spheres, and the three cannot be cleanly divided from the public sphere or even, increasingly, our private spheres. Calls for high-tech ethics, technical experts anointing themselves moral experts, medical codes of conduct, the replication crisis in the psychological and life sciences and its remedies, and the growing concern about public trust in science—in experts—illustrate the intersections among these spheres. Technical expertise cannot be removed from the moral sphere due to the expert relational, social status vis-à-vis the nonexperts.

Approaching this argument from a rhetorical vantage, the moral valance of phronesis is central to questions of goodwill and trust in relationships, cultivated through our rhetorical means, among experts and nonexperts. For Aristotle, phronesis was a central requirement of the good political leader. Today, in a world transformed by techno-scientific modernity, experts take on a vital role in public deliberation, inclusive of the political but also in our social and personal spheres. In the sciences that seem far removed from the day-to-day operations of a community, those experts make decisions about our future, about what research matters, and about what problems are worth raising and solving. The role experts take in public debates, however, is inconsistent, sometimes relying on an authoritarian or paternalistic stance, rather than a collaborative one or, even, a stance that artificially removes the ethical from the epistemic. Such rhetorical moves are, in fact, antithetical to the relational reality that constitutes expert status and the very function of expertise. Expertise without phronesis, in the account I have offered, is a kind of *empeiria*, a knack, insofar as the so-called experts can give no account of the moral realm that their interventions shape, which means their response to a problem is artificially limited to its technical features, which are themselves constituted by historical and social conditions. Kenneth Burke[6] diagnosed the problem in *A Rhetoric of Motives* (1969, 30–31, emphasis in the original):

> If the technical expert, as such, is assigned the task of perfecting new pow-
> ers of chemical, bacteriological, or atomic destruction, his morality *as*
> *technical expert* requires only that he apply himself to his task as effectively
> as possible. The question of what the new force might mean, as released
> into a social texture emotionally and intellectually unfit to control it, or as
> surrendered to men whose *specialty* is *professional killing*—well, that is
> simply "none of his business," as specialist, however great may be his misgiv-
> ings as father of a family, or as citizen of his nation and of the world. The
> extreme division of labor under late capitalist liberalism having made
> dispersion the norm and having transformed the state of Babel into an
> ideal, the true liberal must view almost as an affront the Rhetorical concern
> with identifications whereby the principles of a specialty cannot be taken
> on their face, simply as the motives proper to that specialty. They *are* the
> motives proper to the specialty *as such*, but not to the specialty as *participant*
> *in a wider context of motives.*

Burke's distinction between *morality as technical expert* and *morality as citizen* here illustrates how individuals *might* understand themselves to be performing ade-quately their technical roles despite lacking the broader moral comportment required of them as citizens. Most concerning are those cases where technical experts either fail to see how their work has those broader forces or mistake their morality as technical expert as a substitute for morality as a citizen. Burke notes that "one's morality as a specialist cannot be allowed to do duty for one's morality as a citizen. Insofar as the two roles are at odds, a specialty at the service of sinister interests will itself become sinister" (31). For Burke it was the bomb, but today the consequences of morality as technical expert substituting for morality as citizen can be seen in a wide range of technologies. Consider, for instance, surveil-lance technologies or artificial intelligence that replicates racism and sexism. Consider also those sinister cases in the energy sector which show that scientists and technical persons knew for decades the consequences of their technologies on the exacerbation of anthropogenic climate change. Compartmentalization of morality in scientific fields or technical industries fails because the compartmen-talization of these sectors from public life fails. By design scientific and technical industries have influence over public life yet have minimized their responsibility to participate democratically in that sphere. Phronesis may not redress broader problems here, but it can help to re-center the need for deliberation and judgment about what is good for the community, and not merely expedient, adequate, efficient, or allowed by technical experts. Further, consideration of phronesis and its social constitution demands consideration of and attention to those voices not already included, and underscores the important work of listening. This, too, is important to expertizing itself. Consider Walwema's (2020, 39) study of the World Health Organization's (WHO) Health Alert that provided information about COVID-19 to multiple regions, in multiple languages, with emojis, and with cultural appropriateness, and allowed expert communications to be shared with a much larger audience than any individual expert could reach, and, further, "countered governmental responses that were initially characterized by public denials and reluctance to admit that anything was wrong and that minimized the severity of the outbreak." Indeed, the COVID-19 pandemic has underscored the important roles not only for communications but, as Walwema demonstrates, ethically crafted communication. Such communications about science and health are critically important to COVID-19, but also to the climate crisis, global conversions about inequity, and to combatting growing issues related to mis- and disinformation and other illicit genres (Bojsen-Møller et al. 2020).

The risk society that Ulrich Beck described has come of age and, now, the public has significant stakes in techno-scientific decision-making. Further, experts have a history of failures to engage many publics seriously and morally in deliberations about these issues. Urgency for this call to a moral understanding of expertise can be found across sectors, from academia to industry to government. Distrust of experts should not surprise anyone who has even a cursory understanding of the history of science, the failures of governments to protect their most vulnerable, and the growing disregard corporations have for the well-being of the communities in which they are embedded. Failure to address the moral comportment of experts will mark a failure to address climate change, a failure to stop the tech sector from shaping the lives of millions without consultation, a failure to compel governments to protect their constituents from corporate interests, et cetera. This is not to say experts do not currently have a moral stance toward publics. Rather, this is to say that the rhetorical crafting of the distant, objective expert is often not effective because it removes important relational aspects, including those that build trust. Expert status, once again, is relational and, being so, is beholden to the values and norms of the communities within which expert status is embedded and, indeed, by which it is sanctioned. Not only is expert status governed by the need for a moral comportment, but expertise itself is, too. Expertise requires knowing-that (the theorical knowledge) and knowing-how (the practical knowledge of doing), to be sure, but also, critically, knowing-why (practical moral knowledge as a relational concept). Phronesis is required insofar as expertise requires good deliberation for good decision-making. Good decision-making, in the case of expert knowledge, is relational and thus measured in terms of its attunement to the values and norms of the situation in which the decision-making occurs. Situations are partially constituted by the audience, and, thus, even the enactment of expertise itself requires knowledge of and goodwill toward the audience. The requisite excellence of character or habituation of virtues allows experts to enact their knowledge to address some situation, and therein lies the activity of expertise. Their knowledge is inclusive of relevant theoretical principles, practical skills, and, crucially, the kind of understanding built from experience, memory, and awareness of one's limited perspective, and a habituation of virtue that is described by the idea of phronesis. Phronesis is sometimes overlooked as a key relational, social, and rhetorical concept central to understanding expert status and expertise. In this erasure of expertise's phronetic quality, specialization, technical proficiency, and other seemingly coterminous concepts have obfuscated the most necessary rhetorical features of experts and, in turn, undermined public trust. To redress this erasure, experts, especially those of the technical variety, must reattune themselves to the practical and relational moral qualities of expertise and expert status as a kind of enactment of practical knowledge as phronesis. Cleaving off phronesis from the other forms of knowledge that experts require has left us with an impoverished understanding of expertise.

Perhaps, too, a dangerous understanding of expertise. Collins, Evans, and Weinel (2017) struggle with the legacy of science, technology, and society (STS) studies in what they, in response to Sismondo (2017), diagnose as a "post-truth" era. It is the culpability and the intellectual agreement of STS and this post-truth refrain with which they attempt to reckon. Phronesis, and a rhetorical understanding of expertise, might productively contribute to this important consideration. Rhetoric keys into communication, to invention and argument, to conceptual and preconceptual modes of reasoning, and to the social-historical milieu where our discourses are invoked and unfold. Importantly, rhetoric is also attendant to bad faith appeals, propaganda, and also the ways in which rhetorical-situational developments can enact violence upon us (see, for discussion of the former, Mercieca 2020; and on the latter point, Bernard-Donals 2020). Phronesis provides a broad concept to explore the dimensions of expertise required to address indefensible breaches of trust, including with publics, for instance, which have concerned STS scholars. The concept can accomplish this through an articulation of its relational comportment, which signals duties to others in social and political senses by grounding the obligation in said comportment and tracing its enactment through rhetorical efforts. As an aspect of expertise, too, this concept helps illustrate how ethical decisions are always implicated and socially constituted even among experts, in a varieties of forms, from interpersonal relational aspects to social roles, and, critically, in the conceptions of honesty, truth, and "objectivity" that are frequently constitutive of discourses surrounding the merits of scientific method. More directly, you cannot have *good* science without talking about what "good" means, including who is excluded in current conceptualizations and who may be in proleptic imaginings as they open or restrict possible horizons. Attending to how such conceptions operate, however, does not remove the requisite theoretical and practical knowledge—as, for instance, episteme and techne—that are also constitutive of our conception of expertise and implicated in our assessments of expert status. The good, like phronesis, remains here, like our scientific knowledge, provisional and requires reflexive engagements. Rhetoric, then, offers "a rational way of coming to terms with the provisionality of reason" (Blumenberg 2020, 203).

Notes

Introduction

1. For a discussion of trends in the United States, see Motta (2017) in the *Washington Post*; for large-scale survey data on this matter in terms of institutional trust in the United States, see Pew Research Center (2017).

2. Epstein (2019), however, offers numerous examples of how differing experiences can, in fact, be central to the kinds of expertise humans excel at, including work in changing environments where what we might call a kind of creative thinking is central to high-level performance. See especially the discussion from page 32 on.

3. See, for instance, Hofstadter (1963) and Kakutani (2018).

4. Here an early modern refrain on expertise is recalled. "Expert" was commonly used in the sixteenth century as an adjective to describe an individual's capacities, not their status. "Expert," as a noun, developed later, and "expertise" is a nineteenth-century invention in vocabulary (Ash 2010, 4).

5. See the 2006 Kennedy edition of Aristotle's *Rhetoric*.

6. Later forms and models characterizing various "expertises," such as interactional experts and contributory experts described by Collins and Evans (2007), will be discussed.

7. Walton (1992) provides a distinction between forms of authority, calling them "cognitive" and "administrative." Miller (2003, 186) takes up this distinct and suggests both forms of authority may be deployed in risk assessment debate and deliberation. König (2017, 26) notes that, in ancient cultures, expertise and authority "did not always go together straightforwardly."

8. Although I am going to bracket the discussion of the possibility of "evil experts" as outside the scope of this book, there is reason to consider this question, and I believe authority and trust are key concepts for such an investigation. The concept of evil is, itself, meritorious of study and, rightfully, deployed with caution. However, the history of science and technology demonstrates many evils or vices that we cannot ignore, and experts, too, are at the heart of these stories. While I certainly do not deny that someone who has expertise can engage in immoral or wrong behavior, I do suspect they over-rely on authority in place of expertise and are antidemocratic (that is, as regards including the broad public in deliberations) in their thinking, in opposition to the relational comportment I argue can be facilitated by a virtue-based conception of expertise. Indeed, we daily see unjust acts—including racism, ableism, gender-based discrimination, and other forms of prejudice—sometimes sanctioned by figures of authority or even those who purport to be expert. We also know the moral failures of our communities are replicated in our science, our technology creation, and our application of science in health, for instance. It is, too, replicated in expertizing. These failures, in addition to being moral losses, also hinder progress in understanding the very scientific subjects studied and must be fought on an ongoing basis.

9. I will not rehearse the arguments here for why this is so because it is rather the distinction between authority and expertise that is important. Although some experts may transform their ethos into logos-centric expertise to claim broader authority, it is ultimately this disingenuous transformation that fails because publics-as-audience are capable of assessing—and likely

expect—those. *ethotic* appeals that are marked by arête, enuoia, and phronesis. The sense of logos-centric here is distinct from Hartelius's (2020) conception of logos, taken up below, which is grounded in the thinking of Heraclitus.

10. Deliberations, debates, and so forth, inclusive of and not "augmented by" experts, should involve a kind of civil comportment or disposition (following Keith and Danisch 2020, 73–75). Indeed, such a civil comportment of the kind of "strong civility" Keith and Danisch describe is a critical habit as numerous subjects relevant to expert discourses encounter divisive public debate (e.g., vaccines and climate change); see Danisch and Keith (2020), especially pages 163–70; see also Cagle (2018).

11. Several decades of rhetorical studies committed to the study of science have provided important groundwork for the study of expertise. Investigating how scientific knowledge is produced through a rhetorical lens has demonstrated the importance of communication, language, discourse, debate, and deliberation in scientific discovery and invention, methodological development, and public dissemination and policy deliberation. Within this research, several touchstone studies appeared, focused on questions of expertise. Preceding all these studies, Isocrates, Plato, Aristotle, Cicero, and many other rhetoricians of Greek and Roman antiquity explored questions of specialized knowledge and questions of expertise.

12. Tindale (2012) explains a rhetorical account of expertise necessarily addresses how an audience experiences the appeals made by an expert. An audience is not a passive observer of such appeals, but in fact works with the expert and their appeals to co-construct the message, thus playing a vital role in the success or persuasiveness of an appeal. Further, Tindale argues, using the example of anti-vaccine advocates, that audiences do not reason as individuals, but within communities, sharing values and beliefs. When experts address such audiences, including individual representatives, they must acknowledge and respond to those values and beliefs, not simply ignore them, or else risk having their message not be heard.

13. By "cognitive," I do not mean to suggest the kind of research tradition represented by the psychological sciences per se. Rather, here I mean an understanding of how one can cultivate the habits of the mind through rhetorical training to perform at expert levels. In this way, the rationale for a cognitive approach does not depart from a traditional pragmatic rhetorical approach. Further, it does not suggest limitations on the variety of ways one might cultivate these habits in different types of minds, with different affinities, and offering different ways of being expert. Here, too, the enormous scientific and philosophical evolution of what is meant by the soul, mind, et cetera, should be noted, and the anachronistic conflation of terms are, rather, merely analogical conveniences. See, on the science and philosophy of the Hellenistic period, and with respect to Hellenistic uptake of Aristotle, Annas (1992).

14. Aristotle, it is often noted, believed women to be inferior to men. This is not a surprising or, sadly, uncommon view, from antiquity to today, with some exceptions and permutations of a perhaps "proto-feminism," including the work of some Cynics (e.g., Hipparchia) and the Stoics' human-based conception of virtue (see Bergès 2015, 25, 23). Although the treatment of virtue ethics here is partially grounded in Aristotle's work, especially owing to his influence on rhetorical studies, moral philosophers have provided insightful commentaries and alternatives that are also crucial to the concepts of moral virtues as developed in this book. Echoing Self (1979, 144n4), "I find the assumption that *phronesis* (prudence) is a virtue of only one gender both unfortunate and irrational." Further to the point, on the understanding of his concepts myself, as someone woman-identified, as Sayers (2005, 168) notes, there is "nothing in my shape or bodily functions which need prevent my knowing about him [i.e., Aristotle]." Aristotle also held ableist ideas and disability scholars have done important work to recover virtue ethics from the harmful arguments Aristotle set out (see, e.g., Clifton 2018). Once again, with Aristotle's perspectives, we are starkly reminded that this concept does not provide any

kind of "objective" or "universal" ethical knowledge, but rather tells us about a fallible human capacity. Aristotle's views here underscore the importance of ongoing reflection on how sexism, racism, antisemitism, ableism, transphobia, and a variety of other prejudices cause significant harm. As Katz (1993, 44) writes, "Aristotle is merely reporting the faults of his and every age," insofar as his is an empirical philosophical vantage that reveals the beliefs then held. Notably, too, central as Aristotle is to the theorization of virtue ethics, his thinking on key features can be traced to earlier thinkers, including Plato (see *Gorgias* and *Republic* on emotional harmony and *Laws* on habituation), and were likely more widespread; thus, the danger of "'figurehead' history" should be noted (Bergès 2015, 18; see, on the discussion of Plato's influence, 12–22 especially).

15. Mailloux notes numerous translations of phronesis, including "prudential, ta'aqqul, la prudence, prudence, *praktische Wissen*, practical reason or knowledge" (2004, 457, emphasis in the original). Notably, however, these terms do not always mark the ethical conception offered in Aristotelian thinking, particularly more pragmatic sensibilities about prudence in pre-Ciceronian Rome and in many modern uses of prudence, often reflecting a self-interested manner of proceeding (see Sloane 2001).

16. It is the configuration of society, what Ulrich Beck described as a "risk society," that opens a space for technocratic models to overtake democratic models, as the world we live in requires that we understand increasingly complex technoscientific topics in the course of our daily lives. Beck's (1992) risk society is a sociological model where economic principals of scarcity are augmented by an additional condition, the distribution of risk. Risks in modernity take on a distinct nature due to their qualities of imperceptibility and technoscientific underpinnings. Also, risks are inequitably distributed, which means that there will be global consequences to how risks are distributed, and they will, often, disproportionally affect already marginalized communities. Nuclear energy is an often discussed risk born of modernity, promising better, cleaner energy, but also with its imperceptible but deadly contamination in the event of disasters and even long-term storage of spent fuel rods. Questions of nuclear energy provide a good example of how risk society operates, as it has been a prominent locus for debate surrounding expert and public interactions. Nuclear energy is a complex science and engineering effort that requires considerable expertise. The potential consequences of nuclear disasters, however, are not constrained to technical domains (see Weinberg 1992).

17. Ash (2010) notes that, although questions of expertise are often theorized with cases from modernity, forms of expert knowing were crucial to the development of modernity that took root in the early modern period. Although there are notable distinctions in how expert knowing and sanctioning were conceived, the thread to the early modern period provides insights into how the very idea of expertise has developed.

18. Uncertainty is the long-held domain for a rhetoric predicated on Aristotelian contingency. On the question of uncertainty and rhetorical studies of science, see Ceccarelli (2011) on manufacturing controversy, Walker and Walsh (2012) on inventional potential, and, on the matter of uncertainty in so-called scientific controversy, Graham and Walsh (2019).

19. Or what Walker and Walsh (2012, 10) call an "ethical myth."

20. This model of trying to stuff just the "right" technical ideas into someone's head has spectacularly failed, and those failures broadly can be found under the unsuccessful banner of "deficit model" science communication. More fundamentally even, much of what is claimed as objective knowledge in fact often makes some claim to domains quite outside this imagined degree-zero technical knowledge, telling us rather about values. Weinberg's (1972, 1992) conception of trans-science begins to explain this idea, citing the example of determining the risk of a nuclear generation site. Eventually, quite public decisions must be made about what risks are acceptable and what are not, even so early as to determine how far experts ought to go in

their assessment of risk, noting the expenses and time associated even with that most primary activity. In research activities, too, the complexity of the scientific work further requires significant argument, and indeed justification, itself (see Harris 2018, 2019). Danisch explains that it is in the post-Enlightenment rationality, where experts themselves are seated to debate and steeped in argumentation, that rhetoric is renewed—rhetoric is, indeed, demanded—in this risk society.

21. Phronesis has garnered attention in several fields, including educational research (see Kinsella and Pitman 2012) and popular treatment (see Schwartz and Sharpe 2010), and as a practical form of knowledge or reason (Robson 2019).

22. For a brief discussion on the rejection of Aristotelian *telos* by neo-Aristotelian virtue ethicists, see Gallagher (2018).

23. Here I wish to note a strong caveat on how virtue ethics is treated here: I read this tradition through my own comportment as a rhetorician, not a philosopher. For an account of virtue ethics in philosophy, see Annas (2011).

24. The often recounted three schools include, alongside virtue ethics, deontological and consequentialist perspectives. Deontological considerations are concern with duties or rules and consequentialist thinking is, as the term suggests, concerned with the consequences of one's actions. One can find aspects of a virtue ethics in both and, further, differing perspectives on virtue ethics are gathered under this designation; see, especially, Nussbaum (1999).

25. Further, to be virtuous, there is something Aristotle called the unification of virtue. To be virtuous one cannot simply have one or two of the virtues, but must broadly have command of all virtues, and this is grounded in the idea that phronesis provides the capacities for appropriate deliberation and judgment overall. We do not here need to commit ourselves to this view, but it is instructive in thinking about the complex ways that virtues interact and how they are situationally dependent. Annas (2011) compares this to the skills of a pianist, explaining that we do not expect them to only develop, for instance, the skill of fingering, or, in another instance, only the skill of tempo, but rather we expect an integration of these skills. It is this integration of skills that might constitute what we call a skilled player, an expert pianist, even (see Annas 2011, 87). A problem one might identify with this notion of the unification of the virtues is differences among people's lives and circumstances. With varying circumstances, we might anticipate, there are distinct virtues central to one's life. Unification of virtues, however, does not require enactments of virtues generally. A virtue ethics approach acknowledges that different people lead their lives differently, and indeed are born into diverse kinds of lives. "Virtue is not the kind of thing that can be specified in advance so as to be one size that fits all, precisely because practical intelligence gets things right in very diverse circumstances," Annas explains (95). We must, then, in a situation be capable of acting with the appropriate virtues in the appropriate balance, and this will be dependent on the situation.

26. See also Duffy (2017), especially on the "vices of virtue"; and Duffy (2019), especially on the possible limitations of virtue ethics in rhetoric and writing instruction.

27. Indeed, in absence of others, there is not a possibility for moral perfection, Maimonides explains through *ḥokmah* (the third variety of wisdom outlined at the opening of 3.54): "Imagine a person being alone, and having no connexion whatever with any other person, all his good moral principles are at rest, they are not required, and given man no perfection whatever. These principles are only necessary and useful when a man comes in contact with others" (3.54). On Maimonides's use of phronesis, see Metzger (2014). Metzger argues that Maimonides's conception of phronesis can also be understood as an *internal* rhetoric, theorized through the relationship of intellection and speech, and that allows one to "test our understanding of the relationship between the external and internal dimensions of human experience," in contrast with Aristotle's more external uses (125).

28. Disciplinary traditions, Gobet (2016, 2) suggests, seem to align their inquiry with either knowing-that (sociology, law, philosophy) or knowing-how (psychology, neuroscience, education), but this rather hinders research on expertise and thus an integrated model ought to be pursued. Rhetoric offers a mode of inquiry with some experience in such debates, and some tradition of exploring the relationship between theory and *praxis*, each term a multivariate composite in the rhetorical traditions.

29. Here I mean the social sense of Maimonides rather than the political sense of Aristotle.

30. Cf. Miller 1984 on typification in response to recurrent rhetorical situations. Notably, situation is not merely a temporally constrained event; rather, it is a more complex phenomena embedded within historical, cultural, and social traditions, and shaped, further, by the particular actors, inclusive but not limited to the audience, rhetor, as well as material conditions.

31. The University of Waterloo Office of Research Ethics Approved this research (ORE no. 30589).

32. "Multidisciplinary" describes teams comprising experts from several different domains bringing their specialist knowledge to bear on a problem. "Interdisciplinary" normally describes an approach where an expert incorporates specialist knowledge from another field into their own, a kind of integration of fields. "Cross-disciplinary" may involve drawing from other disciplines without such integration. "Transdisciplinary" research transcends the disciplines to offer a more practical solution to problems. See Mehlenbacher (2009) on these distinctions.

33. Some research lends itself to multidisciplinary research models, and those are the teams I am especially interested in examining. Multidisciplinary teams offer an important site of study for expertise because they bring together several different varieties of expertise in fields and what we might call forms of expertise. In such situations, following Collins and Evans's (2007) taxonomy, for example, we might find interactional expertise, contributory expertise, referred expertise, and a range of meta-expertises used to assess others' expertise and claims to expert status. Those teams that might more artificially compose a team, by encouragement of a grant funding model, for instance, are likely to offer distinct social dynamics that are not theorized here.

34. For a detailed account of the project's methods and rationale, see Moriarty et al. (2019).

35. We will later learn from interviews that citizen scientists sometimes hold degrees in science but in many cases do not, and, as citizen scientists, they often work outside of their area of professional expertise strictly defined.

36. Some participants also agreed to interviews or did not complete the survey but agreed to an interview. These data were not linked, so someone who completed the survey may or may not have completed an interview, and vice versa. Since the survey responses and interviews ask different questions, this does not pose any significant limitation to the study but rather allows for a broad range of rich responses.

37. Because this research program is not concerned with assessing someone's expertise as measured by their performance, but rather learning how expertise is rhetorically described, defined, and assessed by people who work in teams that commonly understand expertise and expert status as important in their work, self-identification is a suitable manner to establish inclusion of participants.

38. Experts interviewed range from junior to senior scholars, represent a wide range of standard demographic measures, come from varied geographical locations, and cover many areas of expertise. Demographic data, however, were not collected because, for the purposes of this qualitative study, trends are not reported, and the surveys and interviews are not meant to represent any kind of statistically driven data.

39. Interviews were conducted over a period of several years by my research assistants and myself, and normally lasted about thirty minutes. Interviews were then reviewed for identifying information, transcribed, and verified for their accuracy. These transcriptions do not normally include discourse particles ("uh," "like," etc.) and have been lightly edited for readability (e.g., adding commas, removing interrupted thoughts such as "and I think—yeah," correcting typos introduced by transcription, and minor changes for clarity) but do not substantially alter the meaning of the participants' responses. We used several versions of NVivo to code transcriptions for a range of projects associated with this study, and for the purposes of reporting responses in this book. Pseudonyms have been generated for interview participants, and all participants are referred to by they/them pronouns as we did not ask for gender identification as part of this research.

40. Some of the questions overlapped and others were designed to learn more about the socialization of each group. Citizen scientists, for instance, may have no formal training in a given area of research—or at all. Conversely, they may hold a doctoral degree in a scientific discipline, but not the same field where they participate in citizen science—and accordingly, more open-ended questions about how they obtained their knowledge or skills were appropriate.

Chapter 1

1. Here attention is given specifically to ethics in humans, but this is not to discount the importance of nonhuman ethics. For a meta-analysis of human and nonhuman ethics in professional communication using actor-network theory, see Roundtree (2020); see also, for a case of human and nonhuman rhetorics and ethics, Clary-Lemon (2020).

2. The vegetative has capacities for growth, such as in plants, but not rationality. The appetitive might be governed by reason but is itself without capacities for complete rationality, and it is this faculty that is responsible for the moral virtues including temperance and courage. The rational faculty of the soul is responsible for the reasoning capacities of humans.

3. Gobet (2016, 218), in his summary of the relevance of the *Nicomachean Ethics* to expertise, aligns what Aristotle meant by *episteme* with propositional knowledge, that which is "universal, invariable, and context-independent." However, Ostwald further reminds us that *episteme* in the model of the intellectual virtues does not simply mean a nonspecific form of knowledge, and confusing matters, Aristotle indeed sometimes uses the term *episteme* in this common, "loose way for knowledge of any kind" (307).

4. Deliberation here is treated in both the sense of the technical requirements of book 2 of the *Nicomachean Ethics* as well as the moral requirements elaborated on in book 6. Hursthouse (2006, 300–307) explains the misapprehension that book 2 simply treated means-ends deliberation and book 6 moved onto rule-case deliberation. To do what is right, for example, when a child falls into the river requires not only moral knowledge but also a kind of technical, practical knowledge. Should one first leap in or run down the bank to get ahead of the child? What Aristotle would call the naturally virtuous person may leap right into the river while the person who has cultivated their virtue may decide, with their experience and technical knowledge, to first outpace the child by running down the riverbank, Hursthouse explains (302–6).

5. Aristotle is quite clear that cleverness (*deinos*) is not the sole capacity (*dynamis*) that measures phronesis, as the ethical dimension is crucial to the concept of practical wisdom: "We must stop for a moment to make this point clear. There exists a capacity [*dynamis*] called 'cleverness,' which is the power to perform those steps which are conducive to a goal we have

set for ourselves and to attain that goal. If the goal is noble, cleverness deserves praise; if the goal is base, cleverness is knavery. That is why men of practical wisdom are often described as 'clever' and 'knavish.' But in fact this capacity <alone> is not practical wisdom, although practical wisdom does not exist without it.... But whatever the true end may be, only a good man can judge it correctly. For wickedness distorts and causes us to be completely mistaken about the fundamental principles of action" (*EN* 6.1144a20–30). *Deinos* allows one to achieve a particular end with success, but the moral valance of the end and acts will determine if this is good or bad, and the capacity itself is wed to the morals of the actor. For the *phronimos*, the act must be good, and so too the end. The *phronimos* thus has good *deinos* (Sim 2018, 193).

6. Delineating sophistic pre-metaphysics and Aristotelian metaphysics is important. Poulakos (1984, 218) reminds us of the chief role of facts and what can be demonstrated in an Aristotelian rhetoric, charting the distinction with those preceding Aristotle: "Unlike his predecessors, who posit the world as it is not (Sophists) or a world that ought to be (Plato), Aristotle's starting point is the world as it is, in its positive structure and tendencies."

7. On Aristotle's conception of habituation (*ethismos*) to cultivate virtues, see "The Habituation of Character" in Sherman (1989). Annas (2011, 86) makes it clear habituation is not merely routine, which is "dangerous" for our ethical living, and I would argue that this distinction can help us understand the routine leading to acceptable levels of performance versus the benefits of habituation in learning, leading to expert capacities. Distinguishing my usage of the term "habituating," further, I do not mean the sense often used in psychological sciences, which refers to the decrease in response to stimuli that an individual experiences. One example of this is when an individual so routinely does not wear their seatbelt while driving that they no longer notice the car's seatbelt reminder system, such as a dinging sound. I do not, however, mean to suggest there are no commonalities between my usage, denoting the unconscious elements of this cognitive process, but rather wish to keep the intentionality of habituating in mind. (Thanks to Devon Moriarty for this helpful distinction and example.)

8. Further to this matter, Toulmin (2001) charts the way that Enlightenment thinking—a search for rational, generalizable or universal, theoretical accounts of complex, particular human activities—inappropriately maps the commitments of natural sciences onto the humanities and social sciences. It is, then, the moral comportment of experts that I wish to explore as a central concept in a rhetorical understanding of experts. In doing so, a poverty of theoretical and practical understandings emerge as reductive models of rationality are applied in place of reasonableness. Of the theoretical approaches to ethics and morality, Toulmin reminds us, that Aristotle formulated theoretical sciences as reasoning from certain, general principles to understand the uncertain, practical cases that need resolution; this, however, is the inverse of everyday reasoning for particular situations, where we, instead, "are more certain about the rights and wrongs of particular cases than about the general principles we appeal to in explaining them" (Toulmin 2001, 136). See as an alternative practice-based model to natural science–based social sciences Flyvbjerg's "*phronetic* social science" (2001; see also Flyvbjerg, Landman, and Schram 2012).

9. Foot (1978, 6) explains that practical wisdom must be accessible to everyone, in her exploration of wisdom as an intellectual or moral virtue, arguing that such wisdom must fall into the latter to be accessible to everyone; or, for what knowledge is required of practical wisdom, it "consists of knowledge which anyone can gain in the course of an ordinary life, is available to anyone who really wants it."

10. On Aristotelian ethos, see Halloran (1982) and Benoit (1990); see also Gross (2006).

11. For numerous examples, see Khan (2005).

12. Ash (2010, 22) argues that episteme and techne lose their stricter Aristotelian distinctions during the early modern period with the rise of natural philosophers.

13. For further account of virtue ethics, see MacIntyre (1981), especially on the necessity of virtue ethics as an antidote to several problematic positions partially born of the Enlightenment; and for a useful introduction, see also Annas (2011).

14. Which is not to slip into deontology per se, but rather to focus on the character of the *phronimos*, as one normally in possession of phronesis, and not restrained to a particular imperative or duty.

15. Cf. with *ren* (仁), the Confucian cardinal "virtue," which requires that "one should keep the company of the virtuous" (Ding 2007, 149); see also, on virtues ethics in Confucius and Aristotle, Sim (2007), especially her introductory remarks on comparative work, incommensurability, and translation, based on which I have included quotation marks around "virtue" here.

16. Nor would this person be an exemplar in the model of Zagzebski's *Exemplarist Moral Theory* (2017), which is a departure from Aristotle's model, and rather grounds understanding of virtues in those individuals who illustrate moral excellence. Indeed, an expert as formulated here is chiefly characterized through excellences other than moral, although I argue that moral excellence is important to the epistemic and relational work of experts. Such an exemplarist approach also helps to address charges of Aristotle's circular definition of the person who has phronesis and, indeed, might be a particular productive approach for rhetorical theory.

17. Plato and Aristotle might have rather articulated this as *understanding* rather than *knowledge*; see Zagzebski (1996), 45–46.

18. On the case for virtue ethics in moral philosophy and a normative epistemology in philosophy aligned with such a framework, see Zagzebski's foundational work *Virtues of the Mind* (1996). Although the argument presented here is not situated in philosophy, the chief arguments for a virtue-based understanding of epistemology is important and compelling.

19. Annas (2006) cautions readers of texts from Greek antiquity to mind their modern perceptions of moral knowledge. Unlike more modern notions of moral knowledge, there is not the same prescriptivism in classic conceptions, nor does access to moral knowledge require special epistemic access as moral knowledge is a kind of knowledge apart from others. Such an argument is not so, Annas explains, as in antiquity moral knowledge was understood to be a practical kind of knowledge, akin to the techne or skill we might describe in the constitution of a builder's knowledge, and thus we have access to these ways of knowing. Alignment of moral knowledge with the techne of a builder means that moral knowledge can be understood as a skill, according to both Plato and the Stoics—or following Aristotle, "like skills" (Annas 2006, 287, emphasis in the original). Importantly, skill here should be distinguished from the vernacular use in the current academy as some lesser-than form of knowledge.

20. Collins and his collaborators have a significant body of work on expertise. Among their prolific research on the subject, see Collins (2004, 2011, 2013, 2014, 2017, 2018); Collins and Evans (2014, 2015); Collins and Sanders (2007); Collins and Weinel (2011); Collins, Evans, and Gorman (2007); and Collins, Evans, and Weinel (2016).

21. Importantly, Collins (2018, 352) also argues that assessment of expertise should not include "rightness or efficaciousness as a criterion," which aligns well with the rhetorical discussions of expertise as phronesis.

22. Michael Polanyi's (1966) articulation of *tacit* knowledge is central to Collins and Evans's (2007) work on expertise. Polanyi's work is somewhat outside the debates in philosophy that advance formal (or what might be called explicit or propositional knowledge) knowledge as the principal form of knowledge (Gobet 2016, 219), and describes a form of knowledge that is not immediately available to us for reflection. Citing Gestalt psychology, Polanyi explains that we "may know a physiognomy by integrating our awareness of its particulars without being

able to identify these particulars" and, further, he writes that his own "analysis of knowledge is closely linked to this discovery" (6). Collins (2010, 148) suggests Polanyi's conception of tacit knowledge emerges in a postwar preoccupation with science and, in this world, "Polanyi was tempted to make tacit knowledge into something mystical and inspirational," but such attention focuses on the individual at the cost of the social aspects of acquiring tacit knowledge. Important to our understanding of expertise in this book, Collins in fact goes so far as to say that the "personal aspect of Polanyi's 'personal knowledge,' then, is not knowledge at all, but is the *process of making good judgements* and that arises out of having stores of tacit knowledge," linking this idea of good judgments to ideas of "intuition" (148–49, emphasis mine). The process of making good judgments might be said to function as an aspect of phronesis, which would be a rather different form of knowledge than Collins seems to mean here, although still a form of knowledge insofar as an Aristotelian model allows.

23. Grounded in activity theory, Engeström (2018, 14) writes of the "locus of expertise" that it "resides in object-oriented collective activity systems mediated by cultural instruments," and of its "composition" he tells us "expertise is inherently heterogeneous and increasingly dependent on crossing boundaries, generating hybrids, and forming alliances across contexts and domains."

24. Engeström's implications for what this means clarifies the stakes. He writes that "expertise cannot be meaningfully reduced to individual competency," and of its composition, "there is no universally valid, homogeneous, self-sufficient expertise" (14). Although the case for practice-based understandings of expertise still requires social elements, such as the master trainer, Engeström argues on the matter of learning and expertise that "practice and emulation of established masters alone cannot meet today's demands for transformative expertise" (14).

Chapter 2

1. What would come to be known as the rhetoric of science is well documented in the introductions to Randy Allen Harris's two-part *Landmark Essays on the Rhetoric of Science* (2018, 2019).

2. Wilson's case, ultimately, seems to be a hard one for theorizing the rhetorical dimensions of expertise, as he conflates what we might call a restricted definition of expertise, constrained by domain knowledge and skills, with an expanded definition of expertise, where one's expert status sanctions speaking on a particular topic or range of topics. Here we find a compelling tension, and rhetoricians have, since Lyne and Howe, continued to examine these aspects of experts and their expertise.

3. As an interesting digression, Lyne and Howe (1990, 145) write that Wilson "all but dismisses the humanities and social sciences, as currently practiced, as irrelevant to the modern age." They then quote from *On Human Nature* (1978, 14) as evidence, where Wilson writes, "If human behavior can be reduced and determined to any considerable degree by the laws of biology, then mankind might appear to be less than unique and to that extent dehumanized. Few social scientists and scholars in the humanities are prepared to enter such a conspiracy, let alone surrender any of their territory." In the intervening thirty years since Lyne and Howe's article appeared, and more than forty years since *On Human Nature* was published, the social sciences and humanities have offered a range of responses that might be said to address this call, making a post-human turn, attending to materiality, ontologies, though not entirely reducing human behavior to "scientific laws" and rarely surrendering territory. Indeed, this is important because, as Lyne and Howe demonstrate, it is a rhetoric of the gene that allows Wilson's "expert" status, but this research program is not "controlled by disciplinary criteria" (146). On the subject of Wilson's research, Lyne and Howe caution that "educated people

[perhaps, we might say, rather those educated in a particular subject] have a responsibility to understand the inherent dangers of such naturalizations," and this caution is also relevant to conceptions of how and why some people excel to become experts (148).

4. For example, they claim Wilson's "position is severely constrained" from a "scientific perspective," and that it is "imagery and rhetoric [that] do work that analysis cannot do, implying conclusions for which authors need take no direct responsibility. So, for instance, the strategy of 'speculation' gives Wilson rhetorical license to invoke science while remaining insulated from technical criticism" (145).

5. Aristotle established ethos as one of three modes of proof, the other two being pathos (emotion) and logos (good reason). In scientifically oriented modernity, logos is perhaps among the most credible appeals, invoking conceptions of logic, evidence, quantitative data, and so on. Pathos is rather denigrated, the appeal to our emotions. In each speech, Aristotle would advocate, one must cultivate one's ethos, and not rely on prior reputation or presume its persuasiveness in speech. Conversely, Isocrates's conceptualization of ethos departs from an Aristotelian perspective in that, for him, "ethos is the speaker's prior reputation, developed during life" (Benoit 1990, 258). Benoit stresses this distinction between Aristotle and Isocrates. He explains that while passages in Aristotle "rule out the 'prior reputation' notion of ethos," Isocrates "explicitly juxtaposes the 'argument which is made by a man's life' with 'that which is furnished by words,'" favoring the former explanation of ethos (258, 257). Further, Isocrates forcefully argues that the cultivations of one's character is essential to becoming persuasive, explaining that "the stronger a man's desire to persuade his hearers, the more zealously will he strive to be honourable and to have the esteem of his fellow-citizens" (Isocrates, *Antid.* 278, Norlin ed.).

6. Cf. Miller (2018) on the matter of character. His treatment is a kind of popular philosophical account, situated within the virtue ethics tradition, wherein psychological studies are drawn on as empirical resources to understand how character is enacted.

7. Hartelius (2011, 29–30, 165–66) offers six rhetorical congruities, a non-exhaustive list, that operate as expert rhetorical strategies: association in expert networks; mastery of a skill or demonstration of techne; positioning toward publics for pedagogical purposes (which includes forms of resistance to "instructing the public"); participating in public conversations; crafting of the rhetorical situation to demand expertise; and crafting a longer-term niche for their expertise.

8. C. P. Snow (2012) made the argument in the late 1950s that the humanities and sciences operate, essentially, as two cultures that do not share a language. This, he suggested, hinders contemporary problem-solving. The concept has been much debated and contested (see Gould 2003, especially 89–95), but has certainly been influential. Gould (2003, 91), crucially, reminds us that Snow's arguments emerge from a particular context, what Gould articulates as a "local British phenomenon," to a broader, global phenomenon. Further, Snow had a political argument in his remarks, concerning poverty, driven by "British paternalism," following Gould's characterization, where "expertise" was to be exported globally (91–93). Snow largely recanted his dichotomous view, but the distinction is still invoked today.

9. Barriers to citizen understanding are not necessarily internal to the citizen. Consider, for example, Kinsella's (2001) and Kinsella and Mullen's (2007) research on the Hanford nuclear site downwinders. Citizens engage in considerable effort to learn about a complex subject; have their own experience, local knowledges, and expertise in medical domains related to elevated cancer rates; and have been hindered by both bureaucratic and material challenges. Indeed, citizens in this example work within the norms of science and offer sanctioned information. Kinsella and Mullen write that "downwinders themselves did not typically presume

to contest the establishment's scientists, but in public forums they enthusiastically broadcast the conclusions of other scientists who did" (88).

10. See Waddell (1990).

11. In the fields of science studies allied to the rhetoric of science, researchers have shown that there are knowledges and forms of expertise that exist outside the conventional cadre of regular, professional experts. Here again we see that limiting expertise to a specialization or profession unnecessarily limits our understanding of the concept. Famously, Wynne (1989, 1992) describes Cumbrian sheep farmers who develop certain forms of local knowledge that provide an expertise that can be integrated with professional scientific knowledge to provide a more complete picture of effects of radiation contamination, following the disaster at Chernobyl. Further to this point, it is instructive to ask how public concerns, knowledges, and even expertises are acknowledged and incorporated into the deliberative process. A broader concern, however, is that as hyper-specialization continues unabated, judgment concerning the relative merits of expert claims is difficult. This extends, too, to academic situations where multidisciplinary research approaches are acknowledged as important to solving wicked problems, but the administrative and also epistemic basis for such work remains uncertain. In these cases, a kind of expertise to negotiate different research traditions, discourses, and practices is necessary.

12. This was also noted in interviews with professional experts.

13. It is especially important that such deliberations might proceed without simply subsuming all forms of argument under a technoscientific rationality; see Fisher (1994); Goodnight (1982); and Kinsella (2001, 2004).

14. In rhetorical studies, the work of Ceccarelli (2011) further charts our peril at ignoring the arguments of sanctions experts when bogus experts weigh in to generate controversy (for political or other dubious gains) on science that is, in fact, settled. Several popular books have also discussed how science is undermined by politicians (Levitan 2017), celebrities (Claudfield 2015), and experts themselves (Freedman 2010).

15. Phronesis has garnered attention in a number of fields, including educational research (Kinsella and Pitman 2012), and even popular treatment (Schwartz and Sharpe 2010) and as a practical form of knowledge or reason (Robson 2019).

16. Such a position aligns with virtue ethics models insofar as, like a virtue approach, the "right" action (here Majdik and Keith speak of outcomes) is not a matter of concern but rather is the "virtuous" action (here Majdik and Keith speak of a "dispositio") to which one ought to attend (see Annas 2011, 47–51, especially).

17. This is consistent with the account of phronesis discussed with respect to Aristotelian virtue ethics—although it departs from the focus on phronesis as moral knowledge—as a practical kind of knowledge and, thus, not only argument functions. Mehta, Majdik, and Platt (2012), however, cite both "technical and moral dimensions" of expertise, suggesting there are two appeals to locate expertise (in the case of risk assessment): one deploying what they call an epistemic register, appealing to empirical assessment of risk, and the other, a "*phronetic practice*," appealing to both social as well as moral aspects of risk (2).

18. Majdik and Keith (2011b) offer a further useful articulation of expertise in this light as the ability to respond to a particular problem: "'Expert' and 'expertise' is not bound (simply) to the possession of knowledge, or processes of knowledge acquisition or production, or connections to knowledge networks, but instead flows from problems that require resolutions" and thus offers a "practice-centric view of expertise [that] originates outside the epistemic register; it starts not with bodies of knowledge, but with people in quandaries" (276). Practice-centric here should be distinguished from practice-based models in the psychological sciences

and, instead, understood in the context of rhetorical thinking and the concept of phronesis as in Majdik and Keith (2011a).

19. Tacit knowledge was previously discussed as related to forms of knowledge and is central to the work of Harry Collins and Robert Evans in science and technology studies on specialist expertises.

20. Analogous concerns appear among philosophers of science who, in a study of 299 such scholars, who felt it is important to conduct work outside their field, but identified perceived and actual barriers to such work (see Plaisance et al. 2019).

21. Collins and Evans (2007) challenge the "social embodiment thesis" and, instead, proposed a "minimal embodiment thesis" as an explanatory mechanism for interactional expertise. Where Dreyfus and Dreyfus (1986) argue that a computer cannot gain intelligence, inclusive of natural language, and Wittgenstein (2009) claims that we could not understand the words of a lion were it to speak as it would play different language games, Collins and Evans argue these are largely abstracted arguments that would benefit from more empirical framing. In a minimal embodiment thesis account, embodiment is required insofar as acquisition of a language well enough to be considered socialized within that linguistic community (77–85).

22. In response to DeVasto, Herndl (2016) raises questions about the case selection, which he suggests may be too tame, in that a solution is reasonably well agreed on and understood, in contrast to more "wicked problems," such as climate change. Herndl suggests the role of values might be foregrounded to offer a richer understanding of how a move to multiple ontologies might better enrich our understanding of expertise. DeVasto (2016b, 15) agrees in principle and explains that the case of L'Aquila shows the dynamic relationship between facts and values, elaborating that "as there are different kinds of doings, value and uncertainty are not flat terms either." DeVasto further explains the importance of situated, embodied experience of seismic activity in the case of L'Aquila to explore the role of publics. Conceding that, following Collins and Evans's cautions we cannot call all experience expertise, questions of values, or matters of concern, become important intersecting ontologies. In this case, we might recall Rice's notion of para-expertise to talk about the experiences and doings that DeVasto describes of the publics. We can also look to examples where experience as subjectivity provides critical forms of expertise (see Arduser 2017).

23. For a discussion of the public sphere in relation to the "technical sphere," see Goodnight (1982).

24. Following DeVasto, Graham, and Zamparutti's (2015) argument that a hybrid deliberation involves experts and publics, Pietrucci and Ceccarelli (2018, 101) argue that "scientists have a special responsibility to enact that hybridity in ethical communication in and around that forum to bridge the gap between the technical and public spheres."

Chapter 3

1. I have attempted to consult multiple perspectives in each field, and it is likely I have inadvertently reduced some of the nuances of the arguments between competing perspectives. Perhaps at my own peril, I have also attempted to provide my most generous readings of these fields and the debates within them, seeking connections rather than contentions among those perspectives. Behind this approach is the rationale that, although in many cases the science is unsettled, each side of the debate's insights provide a new vantage to examine questions of expertise or expert status. Indeed, such an approach is important because, in addition to providing different vantages for study, different fields offer different objects of study.

2. Each field has their respective commitments, and Gobet (2016, 238) provides an accounting of distinct fields of research and their emerging themes.

3. Rhetorically focused inquiry might ask how experts define themselves, who we deem to be an expert, how experts identify or create and exigence for their expertise, how experts cultivate their available means, how expertise is enacted, and how expertise is developed.

4. Socrates, in laying the foundations for what he believes constitutes a society, how people need each other, argues that first, no individual may be self-sufficient, and second, that each individual has distinct aptitudes, arguing that "no two of us are born exactly alike. We have different natural aptitudes, which fit us for different jobs" (*Resp.* 2.2.370b, Lee ed.; on self-sufficiency, see 2.2.369b; on aptitudes, see 2.2.370b).

5. Rhetoric has long attended to the particular situation, but also addressed the character and credibility of a speaker. It seems, in the face of critiques social psychologists make about the "fundamental attribution error" of overemphasizing behavior traits and underemphasizing situational factors, rhetoric has an important theoretical vantage to offer. Indeed, rhetoric attends to a complex constriction of character and its virtues that is necessarily political (Aristotle) and, perhaps more fundamentally, social (Maimonides).

6. The field of rhetoric and composition studies, primarily located in English departments in the United States, has participated in analogous debates (see Hayes and Flower 1981; and Flower and Hayes 1981; see also Hayes and Flower 1986; and Hayes et al. 1987 on cognitive models), their counterparts found in social theories of individual writers. Carter (1990) relates the debates directly to questions of expertise, reviewing the now familiar account of how de Groot and later Simon and others began to identify the importance of domain-specific knowledges, suggesting the importance of what Carter calls "local" knowledge, drawing from Geertz (1983), in addition to more global, generalizable knowledge. Carter, however, advances a "pluralistic theory of expertise," drawing from Dreyfus and Dreyfus (1986) and others, where general and local knowledges are both important to writing and composition processes, and can be understood in the context of a novice to expert continuum toward pedagogical ends. On pedagogy and phronesis in rhetoric, see DeLuca (2020).

7. Infeld at this time held a post in applied mathematics at the University of Toronto, where he taught until 1950 upon his return to Poland. Infeld's departure from Canada, however, was besmirched with fears concerning then Communist Poland and secrets concerning nuclear development. Infeld had visited postwar Poland prior to his ultimate departure from Canada, and this, along with Infeld's work with Einstein, provided the Conservative opposition leader George A. Drew with sufficient information to suggest Communist sympathies and, due to Infeld's work with Einstein, a possible threat to national security (Friedland 2013, 398–99). Infeld's position at the university and his, as well as his children's, Canadian citizenship were revoked. Expert status, we are reminded once again, can become fraught when politicians weigh in on complex specialties such as science for political gain. Indeed, Infeld's expertise was not in nuclear weapon development, and University of Toronto leaders posthumously acknowledged their injury by giving Infeld status as a professor emeritus. Infeld and his children's Canadian citizenship were also reinstated (Horn 1999, 210).

8. Socrates comments, "Quantity and quality are therefore more easily produced when a man specializes appropriately on a single job for which he is naturally fitted, and neglects all others" (Plato, *Resp.* 2.2.369b).

9. For instance, what means are available, including understanding what norms and values are important, what specialist knowledge is relevant, and what balance among the two is required to act judiciously in a particular situation.

10. Crosswhite (2013) treats the subject of wisdom and rhetoric, with some account of Buddhist, Hebrew, Christian traditions, as well as those in antiquity. Wisdom as *sophia* as well

as some conception of practical wisdom is charted, offering wisdom as "the knowledge of knowledge that allows one to decide controversies where no simple standard of knowledge is sufficient because the available standards conflict one another," adding, however, that although the role of rhetoric "seems to promise this wisdom—as judicial deliberation, it seems to be this wisdom—and yet it also seems not to deliver" (322).

11. Whether an activity-based model such as Engeström's, or a Latourian actor-network approach, or even a genre theory–based account, the notion of expert is always a complicated assemblage. However, the rhetorical model here does retain a certain unfashionable affinity for individual human agents as the nexus where action-as-performance-of-expertise is generated. Collins and Evans (2007, 7) seem sympathetic to the challenges posed by a flattening of ontologies as proposed in Latourian frameworks, noting that "humans have an ability to develop and maintain complex bodies of tacit knowledge in social groups that is not possessed by non-human entities." For an interesting case study, see Collins's (1974) study of the CO_2 Transversely Excited Atmospheric laser.

12. Psychological models of experts, we will learn, normally offer insight into how the brain seems to work in terms of the acquisition and retention of knowledge and how that knowledge is recalled to perform in what we might call an expert manner. Some of the most foundational empirical studies on expertise, and some of the best replicated and generalizable but also most contentious and debated, emerge from this field. Although the work presented in this book is not experimental, it is considered apropos to such work in psychology. Studies in psychology gesture toward important insights in the two-millennia tradition of rhetorical studies and evidence some of the practices and claims rhetoricians have used in developing their techniques for training. In conversation with one another, rhetoric and psychology might lead us to develop our sensibilities on how to identify experts, train experts, and explore the cognitive underpinnings of expertise. While such an argument might seem overly expansive, rhetoric has in fact concerned itself with similar questions about the psyche of experts and even features a thread of cognitively inflected studies throughout its history. Rather than entertaining a dangerously scientistic ideological orientation to rhetoric, the dialogue between these two fields offers an approach that allows for the exploration of the creative potentials of expertise, as well as consideration of the social and material conditions that help learners.

13. Further, this definition can provide a way to sort experts from top experts: "This definition has the advantage that it can be applied recursively and that we can define a super-expert: somebody whose performance is vastly superior to the majority of experts" (Gobet 2016, 5; see also Gobet 2011).

14. Now cited over nine thousand times, according to Google Scholar, and over three thousand times according to Web of Science.

15. Unifying decades of research on the science of expertise, Anders Ericsson and Robert Pool's *Peak: Secrets from the New Science of Expertise* (2016) offers a popular introduction to the psychology of practice and how expertise entails practice; for another popular account and challenge to this model, cf. Epstein (2019).

16. Cf. Aristotle's on *dunameis* (abilities or powers) that cannot be changed because they are "implanted in us by nature" (*EN* 2.1.20–25; see also Sherman 1989, especially 177).

17. See Hambrick, Campitelli, and Macnamara's edited volume *The Science of Expertise* (2017). Contributions especially relevant to the present work include Wai and Kell (2017) on intelligence in professional expertise; and Macnamara et al. (2017) on their evaluation of deliberate practice and its operationalized terms and evidence supporting claims about its significance. Other chapters treat research on expertise in the neuroscientific and genetic research.

18. Between the time of writing in 2019 and revisions of this book in late 2020, the miniseries *The Queen's Gambit* has popularized chess by charting the fictional story of a gifted young girl

who becomes a chess prodigy. Compelling a narrative as the series offers, a perhaps more compelling true story is often shared in research on chess and expertise. The remarkable story of the Polgár sisters, trained to be chess grandmasters from their birth in an effort designed by their psychologist father, is invoked as evidence of the importance of deliberate practice. For a brief and popularized account of the Polgár sisters, see Ericsson and Pool (2016) or Clear (2018, 113–14), who embeds the story alongside a summary of the experiment's practical insights. Clear, whose book on habit change is directed toward a popular audience, continues the chapter discussing the role of social norms shaping the possibilities for the kind of social environments that presumably allow for the development of prodigies such as the Polgár sisters—or, in more everyday scenarios, less impressive but nonetheless important habit changes. However, Judit Polgár, commenting on *The Queen's Gambit*, notes the social norms outside her family, meaning those in the chess playing community, were not so supportive, including men who violated norms of play such as shaking hands when they were bested by Polgár or who made comments such as "You're good for a girl" (qtd. in Schlagwein 2020). Here the importance of *phronimoi* as a community may align with the modern premise of habit formation and the importance of social norms and community; therefore, the premise Clear offers affirms the cultivation of a community where one can succeed, underscoring the importance of inclusionary work. Epstein (2019, 13) also details some of the important context of the sisters' family and their upbringing in another popular account of expertise; however, he comes to somewhat different conclusions than the deliberate practice model, suggesting that this model works in only "kind" learning environments (Hogarth, Lejarraga, and Soyer 2015) where closely related and repeatable patterns are central to expert performance.

19. In explaining this distinction they write, "Arational behavior, then, refers to action without conscious analytical decomposition and recombination" (36).

20. Mehlenbacher (2013, 191) explains that "wicked problems frequently have difficult-to-identify beginnings and endings, incomplete information about the rules of play, strategies that can succeed in one setting and fail in another setting that looks identical, unpredictable resources (or pieces), players who do not know the rules or follow them (yet they are shareholders in the outcome of the engagement), and no checkmate—ever—unless we define checkmate as a conclusion defined by running out of time or resources."

21. Not ten thousand hours, exactly; however, Gladwell (2008) made this number famous in the second chapter of his best-selling book *Outliers*, but the original study cited (Ericsson, Krampe, and Tesch-Römer 1993), as well as other studies by Ericsson and collaborators, are more precise and caution in their reporting of how many hours are required for mastery in a given domain: "The average for middle-aged violinists is 7,336 hr, which is so close to the average of 7,410 hr for the best young violinists, that the difference is not statistically significant" (380). Further, Macnamara et al. (2017) caution that while ten thousand hours may be a useful pragmatism it is not especially well grounded theoretically. Expert chess players, for example, have become grandmasters with few hours and others far exceed this number and never achieve such status.

22. In *Are We All Scientific Experts Now?* (2014), Harry Collins provides a personal account of how public trust in experts began to erode from the perspective of someone who grew up in postwar Britain and subsequently studied scientific norms. Overall, Collins charts a movement away from trust in scientific experts since the 1960s to the belief that ordinary people, too, have considerable expertise in science. Although some of the failures of science are implicated in this social change, so too are academic conversations. Collins with his longtime collaborator Robert Evans published "The Third Wave of Science Studies: Studies of Expertise and Experience" (2002), which sets out to map "waves" in the field that chart a movement from explaining the mechanisms of sciences' success to the democratization of science (deflating science), and finally to their third-wave study of expertise. Their model, however, has been

challenged on the grounds that it oversimplifies the diversity of thought in the field during what they demarcate as the "second wave"; also controversial were their proposed solutions to redress failures and omissions of thought in the second wave, including "core sets" to limit claims to expert status (Jasanoff 2003; Rip 2003).

23. Matters are somewhat more complicated than this simple translation suggests. Cape (2003) reminds us that, prior to the rediscovery of Aristotle's works in the twelfth century, Cicero had a considerable influence in the West. His *De inventione* and *De officiis* were widely influential and, upon the later discovery of a complete *De oratore* in the early fifteenth century, Cicero's influence on thinking about *prudentia* would continue to have an impact throughout the Renaissance. Indeed, St. Thomas Aquinas, Cape notes, draws primarily from Cicero, with only modest modification on the basis of Aristotle's works, to develop his conception of *prudentia* (36). In Roman antiquity before Cicero, the term *prudentia* was not well defined philosophically, Cape explains, and it was the word *sapientia* that was instead used to mean wisdom in theoretical and practical senses. While there is some debate about whether these terms were commonly used, it may have been Cicero's work that conflates them, in an effort to "elevate the traditional, practical, Roman ideal of the *prudens* to the status of the Greek philosophical ideal of the *sapiens*" (39). Crucial to our discussion here, of phronesis and *prudentia*, Cape explains that "although *prudentia* represented practical wisdom gained from experience, as opposed to theoretical or philosophical wisdom, or *sapientia*, the distinction does not appear to have been clear in everyday usage before Cicero's day" (39). This is noteworthy as to not suggest a simplistic continuity from Aristotle through to the medieval notions discussed in this chapter.

24. *De inventione*, however, does not treat the subject of rhetorical memory (Carruthers 2006, 212). Yet here we see a critical move in linking memory to the virtues.

25. For a brief account of how virtue ethics can be understood, broadly, during the medieval period, see Haldane (2017); notable as well is the way psychological models of this period are markedly distinct from the manner in which we understand psychology now.

26. *Phantasmata* might be understood in the Greek tradition as "representations," "sort-of-pictures," or "kind of eikōn" (in Latin, *simulacrum* or *imago*) (Carruthers 1990, 16, 17). *Phantasmata* are understood to be physically impressed on the memory, with a pervasive metaphor likening memory to writing on a wax tablet, a metaphor that appears across Greek, Roman, and medieval descriptions of memory. However, it is important, Carruthers argues, to not mistake the tenor and vehicle in this metaphor. "Ancient Greek," she writes, "had no verb meaning 'to read' as such: the verb they used, *ànagignoskó*, means 'to know again,' 'to recollect.' It refers to a memory procedure"; further, "the Latin verb used for 'to read' is *lego*, which means literally 'to collect' or 'to gather,' referring also to a memory procedure" (30, emphasis in the original).

27. Memory has been undertheorized in the rhetorical tradition, relative to its counterparts in the canon, notably invention and style, but was treated at length in philosophical and ethical works; in the rhetorical tradition, memory was a practical affair, as the techne for memory offered in *Rhetorica ad Herennium* demonstrates (Carruthers 2006, 209).

28. Vivian (2018) offers a review of rhetorical treatment of memory as it is reinvigorated in contemporary scholarship on collective memory, tracing the common metaphysical commitments and the disjunctions between classical and contemporary treatments of memory.

29. Although Aristotle's work on *memoria* would not be known until the late twelfth century in the West, Carruthers (1990, 46–47) explains that she treats Aristotle's conceptions as representative of thinking on the subject.

30. Carruthers is careful to distinguish visual from pictorial, explaining that it is not simply by pictorial means by which memory is encoded, but includes other visual means, including "written words and numbers, punctuation marks, and blotches of color; if we read music, we can see it as notes on the staff" (18).

31. Contemporary rhetorical inquiry has furthered and, crucially, sharpened our percipience regarding the cognitive attunements of experts in the study of rhetorical figures as a kind of scheme. Figures are often divided into two primary categories: schemes and tropes. Schemes are marked by formal features such as syntactical structure or morpho-lexical arrangement; while tropes are normally indicated by semantic divergence from an imagined degree-zero (Dubois et al. 1981) language (i.e., language that is seemingly unfigured). A growing body of scholarship in rhetoric attends to the logics of figures and the cognitive affinities that rhetorical principles and practices reveal. Providing the foundation for this movement in rhetorical studies is Fahnestock's (1999) research on *Rhetorical Figures in Science*. Fahnestock's careful study of rhetorical figures shows us how experts deploy figures, arguing that they epitomize their arguments, but indeed this shows us that rhetorical figures function as ways of understanding how to represent data points and assemble them in meaningful ways. Others have expanded this work to a more cognitive framework (see Harris and Di Marco 2009; Harris 2013a, 2013b; Gladkova, Di Marco, and Harris 2015, 2016; Yuan 2017). Previously, rhetoricians of have suggested and explored the cognitive underpinning of figures of speech, including Vico, Buck, and others (see Mehlenbacher and Harris 2017). In this way, figures become an important rhetorical concept for understanding how aspects of memory are constrained and afforded. Rhetorical figures mark certain affinities of the mind (see, especially, Harris 2020) and, in doing so, remind us that these modes of thinking are quite common across eras and cultures and not especially constrained solely by a spoken language. Practically this means we can chart some reasonably global mechanisms implicated in memory. Features of these figures, as we will see, such as division, repetition, and so on, are well established in literature on memory. Importantly, rhetorical figures bridge the work of memory to the work of argument. By marking certain cognitive affinities, rhetorical figures mark rhetorical-structural features of how we interpret and store information, but also how we compose said information into knowledge. Here the important work of Jeanne Fahnestock illustrates how figures epitomize arguments, offering a kernel of insight into the minds of notable experts, including Charles Darwin. Rhetorical theory's considerable preoccupation with figures reaches from antiquity to the present day. Features of figures have interested scholars and practitioners, with many devotees to figuration as a flourish, but here I am interested in those who see figures as fundamental to not only communication but also the mind.

32. There are quite critical distinctions between these periods, as well as some misunderstanding concerning the lines of influence from Greek antiquity to the medieval period, particularly concerning the significance and interpretation of the *Rhetorica ad Herennium*. For an important accounting of these distinctions and scholarly misinterpretations, see Carruthers (2006).

33. See especially Gobet (2016, page 40 and onward) for a summary of how ancient memory techniques are verified by psychological research today.

34. It is worth pausing to note that terms commonly used in contemporary research on memory—chunk, schema, retrieval structure—are debated, and words such as "chunk" are sometimes used to mean quite different theories of process (see, for a helpful overview, Gobet, Lane, and Lloyd-Kelly 2015).

35. Emotions are central to a virtue model, as Sherman (1989, 49) reminds us; for Aristotle, virtues are defined with respect to emotions, and "to hit the mean is to act in a way that is appropriate to the case, but equally to respond with the right sort of emotional sensitivity ..., to act in the manner of virtue." Following Nussbaum (1986), Sherman understands the failure vis-à-vis Aristotle's initial dismissal of Socrates and later seeming agreement at the end of book 7 of the *Nicomachean Ethics* as not an intellectual failure but instead motivational as a "failure to see in a way that could motivate" (47–48n60). Rhetorically, it is notable that these

emotions are those we communicate to others and thus social experiences, realized through their rhetorical activities.

36. Pathos, from Aristotle's *De Anima*, is understood to be "what a sense perception causes in the soul as a kind of image, the having of which we call a memory" (Carruthers 1990, 68).

Chapter 4

1. Participants were also asked to complete a Likert-based list of adjectives to indicate how closely they associate a given word with the idea of an expert. What those results show is that a range of forms of knowing are represented, with the common division of episteme and techne, mostly notable in the common use of knowledge and experience are markers of expertise. The list of adjectives and instrument were adapted from Ohanian (1990; see, especially, table 2, page 44).

2. Haru's point is well taken, with the knowledge that some humanities disciplines (including rhetorical studies) do rely on conferences, publications, and the like, to disseminate findings. However, there are certainly differences between fields, even in terms of how quickly publications will reach an audience. Consider the value of proceedings papers in, for instance, computer science, whereas these are often largely ignored in, say, an English department. Variation exists, too, even within fields. For example, while some areas of English studies may not value conference proceedings, other fields, including technical communication, realize the importance and the value of the genre.

3. We asked respondents to (1) express your confidence in your ability to evaluate the expertise of someone inside of your specialization/area of research; (2) express your confidence in your ability to evaluate the expertise of someone outside of your area but within your field (e.g., you are a molecular biologist working with a plant biologist); (3) express your confidence in your ability to evaluate the expertise of someone outside of your field in another science, technology, engineering, or math discipline (e.g., you are a molecular biologist working with a statistician); and (4) express your confidence in your ability to evaluate the expertise of someone outside of science, technology, engineering, or math disciplines (e.g., you are a molecular biologist working with a historian).

4. The matter of being able to communicate or explain one's reasoning or thinking to others is a recurring theme in both survey and interview data. In the survey data, several participants included the words "teacher" or "mentor" as a term they would associate with an expert.

5. Indeed, this is a well-documented problem for teams, often referred to as the "brilliant jerk" phenomenon. See, for popular discussion of the problem, Sutton (2010), especially chapter 6, "The Virtues of Assholes," although the title should not mislead one to think that it reports on virtues in the sense used in this book; for its consequences, Vettese (2019) and Williams and Multhaup (2018); and for its solutions, Todd (2019).

6. No less important in industry, interpersonal relationships shape teams. Participants who work in industry noted this and explained how they still identify or at least, evaluate those team members. In contrast, academia often provides more space for teams to make decisions about how they will be constituted and with whom they will collaborate.

7. It may be difficult to translate this practice into expert-nonexpert discourse, since a common feature of media training is simplifying messages. Thus, the kind of nuance and complexity an expert would bring to a conversation is filtered out by genre norms. However, as we have seen with the COVID-19 pandemic, many experts in public health have demonstrated that it is possible to introduce complexity, such as uncertainty around surface spread, and clear guidelines to mitigate risk (e.g., handwashing after handling a package will help reduce whatever danger may have been present).

Chapter 5

1. Inspiration for this scene is taken from Sharman Apt Russell's engaging book on tiger beetles, *Diary of a Citizen Scientist: Chasing Tiger Beetles and Other New Ways of Engaging the World* (2014).

2. In another version of this kind of top-down citizen science, a researcher may have participants help them collect water sample readings to test, for instance, for cesium in ocean water following a nuclear disaster, which is what the Fukushima InFORM network in Canada did. In these projects, citizen scientists may or may not have any training in the areas in which they are participating. Or someone may have training in one area of science and wish to participate in another as a citizen scientist. For more information on the projects noted, see, on the Galaxy Zoo project, https://www.zooniverse.org; on the Fukushima InFORM project, https://fukushimainform.ca; and for an analysis, Kelly and Maddalena (2016).

3. These sometimes happen in the wake of disaster, such as the responses of Hanford Downwinders (see Kinsella 2001; Kinsella and Mullen 2007) or the Safecast project (see https://safecast.org). In the case of Safecast, after the nuclear disaster at Fukushima Daiichi in 2011, the group began collecting radiation contamination readings and publishing them openly online, and to date they continue their work. These individuals were not sanctioned experts in nuclear contamination monitoring, which is a complicated matter, but they gained enormous proficiency and brought to bear a suite of advanced, expert, technical skills to the problem to design better tools for monitoring. However, technical or scientific skills alone, particularly in the cases of grassroots citizen science, do not always provide a path to expert status. Ottinger (2010) demonstrated how grassroots efforts to monitor air pollution were challenged by sanctioned scientists for not meeting the standards of practice, although the standards were disputed. Rhetorically, the work Safecast undertook to establish credibility was not unlike what one would expect of a scientist, explaining that their data are not political, that they are interested in understanding the problem, providing evidence-based information, and sharing their methods and approach to data collection. Indeed, in matters of sharing process and data, Safecast was quite exemplary in its work. The group developed technologies to measure radiological contamination, building kits that could be mounted on a car or a bike, and even developed a Geiger counter, working with the established manufacturer International Medcom. See, for analysis, Mehlenbacher and Miller (2018).

4. Citizen scientists, too, are sometimes as expert as the most professionalized experts—indeed, even leaders of fields—and social structures are here implicated as well. Consider, for example, Anne Innis Dagg, the "Jane Goodall of giraffes," who was, for decades, excluded from the ranks of tenured faculty. Amidst dubious rejections from her tenure and promotion committee to unprofessional and overtly gendered dismissal for job opportunities, Dagg continued her science apart from institutionally supported labs, professorships, and the like (see Dagg 2016). Dagg, who only learned of the term citizen science in 2014, demonstrates how those who are marginalized and excluded from professional spheres might continue to make foundational contributions to scientific knowledge.

5. Tide pooling is an activity that occurs with low tide. Small pools of water are left behind as the tide goes out, and sea life is exposed for closer observation until the tide again returns.

6. There still remains the question of those experts who might be brilliant jerks. Optimistically, one would hope that such individuals would not be long tolerated. But are these individuals experts? Undoubtedly many of them may have credentials, strong track records, good professional standing, and specific technical or specialized knowledge and skills. Still, we can ask, are they experts? Yes, we can answer, they are experts if expert status merely requires attributional features and perhaps some specialized knowledge or skills. If we take experts to

have some special status, excelling even beyond those who have significant mastery of some knowledge or skill, we can begin to ask more interesting questions about attribution of expertise as well as what experts know. It seems perhaps most apparent that where brilliant jerks are likely to falter as experts is in their lack of humility, which allows for the kind of growth necessary if we understand expertise as a process of becoming rather than static mastery.

7. On the kind of argument required to accomplish such situating as well as a discussion of how this kind of argument is defined by the community of researchers, see Swales (1990, 2004).

8. If you argue instead that one merely is working in a favored area, citing the right circles of researchers, responding only the exigences that journal editors care about, then we do not entirely disagree. Indeed these very questions implicate the some of the kinds of rhetorical work and practical knowledge I assert are central to expertizing.

Conclusion

1. Even chess players operate in a community where certain norms and values shape their expert performance in ways that demand phronesis, as was noted in the discussion of the International Chess Federation's FIDE Code of Ethics.

2. For instance, the Pew Research Center in the United States recently reported on "Trust and Mistrust in Americans' Views of Scientific Experts" and although researchers found the public has confidence in scientists, trust is divided along partisan political lines and also, interestingly, with how familiar one is with scientists and their research. The report explains when asked if participants have "a great deal" of confidence in scientists, 43% of Democrats affirmed while only 27% of Republicans did the same; and when asked if scientists should take on advising roles in policy debates, 73% of Democrats agreed and, conversely, 56% of Republicans "say scientists should focus on establishing sound scientific facts and stay out of such policy debates" (Pew Research Center 2019, 6).

3. In my book *Science Communication Online* (2019, 150), I document some of these events and argue that, in these cases, "driving the response of scientists is that old rhetorical concern of civic discourse. In a moment when all those rhetorical tools of argument, even the most basic modes of persuasion—logos (good reason), and even ethos (credibility) and pathos (emotions, dealing with responsibility)—have fallen aside, replaced by coercion or force in politics, there is promise in the efforts scientists have undertaken to challenge what is not true or just."

4. Millgram (2015, 37) provides the example of premodern medicine, writing that "patients were being tortured and slaughtered, in the ghastliest ways imaginable, for no good reason at all. But no one could tell, because they weren't equipped to assess the theories, inferential practice, and effectiveness of the procedures performed by members of a specializes professional guild"; and, getting to crux of distrust in experts, he notes that "by far the best medical choice anyone could make was to have nothing to do with doctors."

5. Tuskegee refers to the Tuskegee Study of Untreated Syphilis, carried out by the United States Public Health Service on unknowing participants (Solomon 1985). Henrietta Lacks was an African American woman whose cancer cells were taken without her knowledge or consent and have been used in significant medical research (Skloot 2010). Flint here refers to the Flint, Michigan, water crisis that began in 2014 and continues at the time of writing (early 2020), where drinking water has been heavily contaminated with lead.

6. Previously, I noted some of Aristotle's prejudices, which must be accounted for when we engage his thinking, especially on ethical matters. Here, too, we must do so for Burke; concerning his antisemitism, see Fernheimer (2016).

Bibliography

Annas, Julia. 1992. *Hellenistic Philosophy of Mind*. Berkeley: University of California Press.

———. 2006. "Moral Knowledge as Practical Knowledge." In *The Philosophy of Expertise*, edited by Evan Selinger and Robert P. Crease, 280–301. New York: Columbia University Press.

———. 2011. *Intelligent Virtue*. Oxford: Oxford University Press.

Anscombe, Gertrude Elizabeth Margaret. 1958. "Modern Moral Philosophy." *Philosophy* 33 (124): 1–19.

Aquinas, Thomas. 1964–81. *"Summa Theologiae": Latin Text and English Translation, Introductions, Notes, Appendices, and Glossaries*. Blackfriars ed. 61 vols. New York: McGraw-Hill.

Archer, Lauren. 2012. "Evaluating Experts: Understanding Citizen Assessments of Technical Discourse." In *Between Scientists and Citizens*, edited by Jean Goodwin, 53–62. Ames, IA: Great Plains Society for the Study of Argumentation.

Arduser, Lora. 2017. *Living Chronic: Agency and Expertise in the Rhetoric of Diabetes*. Columbus: Ohio State University Press.

Aristotle. 1962. *Nicomachean Ethics*. Translated by Martin Ostwald. Indianapolis: Bobbs-Merrill.

———. 1984a. *Metaphysics*. In *The Complete Works of Aristotle*, edited by Jonathan Barnes, 2:1552–728. Princeton: Princeton University Press.

———. 1984b. *On Memory*. In *The Complete Works of Aristotle*, edited by Jonathan Barnes, 1:714–20. Princeton: Princeton University Press.

———. 1984c. *Posterior Analytics*. In *The Complete Works of Aristotle*, edited by Jonathan Barnes, 1:114–66. Princeton: Princeton University Press.

———. 1991. *On Rhetoric: A Theory of Civic Discourse*. Translated by George A. Kennedy. Oxford: Oxford University Press.

Ash, Eric H. 2010. "Introduction." In *Expertise: Practical Knowledge and the Early Modern State*, edited by Eric H. Ash, 1–24. Chicago: University of Chicago Press.

Beck, Ulrich. 1992. *Risk Society: Towards a New Modernity*. Thousand Oaks, CA: Sage Publications.

———. 2009. *World at Risk*. Cambridge: Polity Press.

Benoit, William. 1990. "Isocrates and Aristotle on Rhetoric." *Rhetoric Society Quarterly* 20 (3): 251–59.

Bereiter, Carl, and Marlene Scardamalia. 1993. *Surpassing Ourselves. An Inquiry into the Nature and Implications of Expertise*. Chicago: Open Court.

Bergès, Sandrine. 2015. *A Feminist Perspective on Virtue Ethics*. London: Palgrave Macmillan.

Berkenkotter, Carol, and Thomas N. Huckin. 1993. "Rethinking Genre from a Sociocognitive Perspective." *Written Communication* 10 (4): 475–509.

———. 1995. *Genre Knowledge in Disciplinary Communication: Cognition/Culture/Power*. Hillsdale, NJ: Lawrence Erlbaum Associates.

Bernard-Donals, Michael. 2020. "On Violence and Vulnerability in a Pandemic." *Philosophy and Rhetoric* 53 (3): 225–31.

Bitzer, Lloyd F. 1968. "The Rhetorical Situation." *Philosophy and Rhetoric* 1 (1): 1–14.

Blumenberg, Hans. 2020. "An Anthropological Approach." In *History, Metaphors, Fables: A Hans Blumenberg Reader*, edited by Hannes Bajohr, Florian Fuchs, and Joe Paul Kroll, 177–208. Ithaca, NY: Cornell University Press. First published 1971.

Bojsen-Møller, Marie, Sune Auken, Amy Devitt, and Tanya Karoli Christensen. 2020. "Illicit Genres: The Case of Threatening Communications." *Sakprosa* 12 (1): 1–53.

Bullard, Robert D. 2018. *Dumping in Dixie: Race, Class, and Environmental Quality*. New York: Routledge. First printed 1990.

Burke, Kenneth. 1969. *A Rhetoric of Motives*. Berkeley: University of California Press.

———. 1984. *Permanence and Change: An Anatomy of Purpose*. Berkeley: University of California Press.

Cagle, Lauren E. 2018. "Climate Change and the Virtue of Civility: Cultivating Productive Deliberation Around Public Scientific Controversy." *Rhetoric Review* 37 (4): 370–79.

Cape, Robert W. 2003. "Cicero and the Development of Prudential Practice at Rome." In *Prudence: Classical Virtue, Postmodern Practice*, edited by Robert Hariman, 35–66. University Park: Penn State University Press.

Carr, David, James Arthur, and Kristján Kristjánsson. 2017. "Varieties of Virtue Ethics: Introduction." In *Varieties of Virtue Ethics*, edited by David Carr, James Arthur, and Kristján Kristjánsson, 1–13. London: Palgrave Macmillan.

Carruthers, Mary J. 1990. *The Book of Memory: A Study of Memory in Medieval Culture*. Cambridge: Cambridge University Press.

———. 2006. "Rhetorical *Memoria* in Commentary and Practice." In *The Rhetoric of Cicero in Its Medieval and Early Renaissance Commentary Tradition*, edited by Virginia Cox and John Ward, 209–37. Leiden: Brill.

Carruthers, Mary, and Jan M. Ziolkowski. 2002. "General Introduction." In *The Medieval Craft of Memory: An Anthology of Texts and Pictures*, edited by Mary Carruthers and Jan M. Ziolkowski, 1–31. Philadelphia: University of Pennsylvania Press.

Carter, Michael. 1990. "The Idea of Expertise: An Exploration of Cognitive and Social Dimensions of Writing." *College Composition and Communication* 41 (3): 265–86.

Caulfield, Timothy. 2015. *Is Gwyneth Paltrow Wrong About Everything?* Boston: Beacon Press.

Ceccarelli, Leah. 2011. "Manufactured Scientific Controversy." *Rhetoric and Public Affairs* 14 (2): 195–228.

Chase, William G., and Herbert A. Simon. 1973. "Perception in Chess." *Cognitive Psychology* 4 (1): 55–81.

Cicero. 1942a. *On the Orator, Books 1–2*. Translated by E. W. Sutton and H. Rackham. Loeb Classical Library. Cambridge, MA: Harvard University Press.

———. 1942b. *On the Orator, Book 3. On Fate. Stoic Paradoxes. Divisions of Oratory*. Translated by H. Rackham. Loeb Classical Library. Cambridge, MA: Harvard University Press.

———. 1949. *On Invention. The Best Kind of Orator. Topics*. Translated by H. M. Hubbell. Loeb Classical Library. Cambridge, MA: Harvard University Press.

Cicero, pseudo. 1964. *Rhetorica ad Herennium*. Translated by Harry Caplan. Loeb Classical Library. Cambridge, MA: Harvard University Press.

Clary-Lemon, Jennifer. 2020. "Examining Material Rhetorics of Species at Risk: Infrastructural Mitigations as Non-Human Arguments." *Enculturation* 32, http://enculturation.net/material_rhetorics_species_at_risk.

Clear, James. 2018. *Atomic Habits*. New York: Penguin.

Clifton, Shane. 2018. *Crippled Grace: Disability, Virtue Ethics, and the Good Life*. Waco, TX: Baylor University Press.

Collins, Harry M. 1974. "The TEA Set: Tacit Knowledge and Scientific Networks." *Science Studies* 4 (2): 165–85.

———. 2004. "Interactional Expertise as a Third Kind of Knowledge." *Phenomenology and the Cognitive Sciences* 3 (2): 125–43.

———. 2010. *Tacit and Explicit Knowledge*. Chicago: University of Chicago Press.

———. 2011. "Language and Practice." *Social Studies of Science* 41 (2): 271–300.

———. 2013. "Three Dimensions of Expertise." *Phenomenology and the Cognitive Sciences* 12 (2): 253–73.

———. 2014. *Are We All Scientific Experts Now?* Cambridge: Polity Press.

———. 2017. *Gravity's Kiss: The Discovery of Gravitational Waves*. Cambridge, MA: MIT Press.

———. 2018. "Are Experts Right or Are They Members of Expert Groups?" *Social Epistemology* 32 (6): 351–57.

Collins, Harry, and Robert Evans. 2002. "The Third Wave of Science Studies: Studies of Expertise and Experience." *Social Studies of Science* 32 (2): 235–96.

———. 2007. *Rethinking Expertise*. Chicago: University of Chicago Press.

———. 2014. "Quantifying the Tacit: The Imitation Game and Social Fluency." *Sociology* 48 (1): 3–19.

———. 2015. "Expertise Revisited, Part I—Interactional Expertise." *Studies in History and Philosophy of Science Part A* 54:113–23.

Collins, Harry, Robert Evans, and Mike Gorman. 2007. "Trading Zones and Interactional Expertise." *Studies in History and Philosophy of Science Part A* 38 (4): 657–66.

Collins, Harry, Robert Evans, and Martin Weinel. 2016. "Expertise Revisited, Part II: Contributory Expertise." *Studies in History and Philosophy of Science Part A* 56:103–10.

———. 2017. "STS as Science or Politics?" *Social Studies of Science* 47 (4): 580–86.

Collins, Harry, and Gary Sanders. 2007. "They Give You the Keys and Say 'Drive It!' Managers, Referred Expertise, and Other Expertises." *Studies in History and Philosophy of Science Part A* 38 (4): 621–41.

Collins, Harry, and Martin Weinel. 2011. "Transmuted Expertise: How Technical Non-Experts Can Assess Experts and Expertise." *Argumentation* 25 (3): 401–13.

Crosswhite, James. 2013. *Deep Rhetoric: Philosophy, Reason, Violence, Justice, Wisdom*. Chicago: University of Chicago Press.

Csikszentmihalyi, Mihaly. 1990. *Flow: The Psychology of Optimal Experience*. New York: Basic Books.

———. 1996. *Creativity: Flow and the Psychology of Discovery and Invention*. New York: HarperCollins.

———. 1997. *Finding Flow: The Psychology of Engagement with Everyday Life*. New York: HarperCollins.

Dagg, Anne Innis. 2016. *Smitten by Giraffe: My Life as a Citizen Scientist*. Kingston, ON: McGill-Queen's University Press.

Danisch, Robert. 2010. "Political Rhetoric in a World Risk Society." *Rhetoric Society Quarterly* 40 (2): 172–92.

de Groot, Adriaan D. 1965. *Thought and Choice in Chess* [*Het denken van den schaker: Een experimenteelpsychologie studie*]. Translated by G. W. Baylor. New York: Basic Books. First published 1946 in Dutch.

DeLuca, Katherine. 2020. "Fostering Phronesis in Digital Rhetorics." In *Digital Ethics: Rhetoric and Responsibility in Online Aggression*, edited by Jessica Reyman and Erika M. Sparby, 231–48. New York: Routledge.

DeVasto, Danielle. 2016a. "Being Expert: L'Aquila and Issues of Inclusion in Science-Policy Decision Making." *Social Epistemology* 30 (4): 372–97.

———. 2016b. "Matters of Concern and the Politics of Who: A Response to Herndl." *Social Epistemology Review and Reply Collective* 5 (7): 14–17.

DeVasto, Danielle, S. Scott Graham, and Louise Zamparutti. 2015. "Stasis and Matters of Concern: The Conviction of the L'Aquila Seven." *Journal of Business and Technical Communication* 30 (2): 131–64.

Ding, Huiling. 2007. "Confucius's Virtue-Centered Rhetoric: A Case Study of Mixed Research Methods in Comparative Rhetoric." *Rhetoric Review* 26 (2): 142–59.

———. 2008. "The Use of Cognitive and Social Apprenticeship to Teach a Disciplinary Genre: Initiation of Graduate Students into NIH Grant Writing." *Written Communication* 25 (1): 3–52.

———. 2009. "Rhetorics of Alternative Media in an Emerging Epidemic: SARS, Censorship, and Extra-Institutional Risk Communication." *Technical Communication Quarterly* 18 (4): 327–50.

Douglas, Heather. 2018. "Resisting the Great Endarkenment: On the Future of Philosophy." *Philosophical Inquiries* 6 (2): 93–106.

Dreyfus, Hubert, and Stuart E. Dreyfus. 1986. *Mind over Machine.* New York: Free Press.

Dubois, Jacques, Francis Edeline, Jean-Marie Klinkenberg, Philippe Minguet, François Pire, and Hadelin Trinon. 1981. *A General Rhetoric [Rhétorique générale].* Translated by Paul B. Burrell and Edgar M. Slotkin. Baltimore: Johns Hopkins University Press. First published 1970 in French.

Duffy, John. 2017. "The Good Writer: Virtue Ethics and the Teaching of Writing." *College English* 79 (3): 229–50.

———. 2019. *Provocations of Virtue: Rhetoric, Ethics, and the Teaching of Writing.* Logan: Utah State University Press.

Duffy, John, John Gallagher, and Steve Holmes. 2018. "Virtue Ethics." *Rhetoric Review* 37 (4): 321–92.

Engeström, Yrjö. 2018. *Expertise in Transition: Expansive Learning in Medical Work.* Cambridge: Cambridge University Press.

Epstein, David. 2019. *Range: Why Generalists Triumph in a Specialized World.* New York: Riverhead Books.

Ericsson, K. Anders, Ralf T. Krampe, and Clemens Tesch-Römer. 1993. "The Role of Deliberate Practice in the Acquisition of Expert Performance." *Psychological Review* 100 (3): 363–406.

Ericsson, K. Anders, and Robert Pool. 2016. *Peak: Secrets from the New Science of Expertise.* Boston: Houghton Mifflin Harcourt.

Fahnestock, Jeanne. 1998. "Accommodating Science: The Rhetorical Life of Scientific Facts." *Written Communication* 15 (3): 330–50. First published 1986.

———. 1999. *Rhetorical Figures in Science.* Oxford: Oxford University Press.

Fan, Lai-Tze. 2020. "Research-Creation for the Community: Pedagogy, Feminist Maker Cultures, and the Critical Work of Making Face Masks in the Time of COVID-19." *ESC: English Studies in Canada* 44 (4): 39–46.

Fernback, Jan, and Zizi Papacharissi. 2007. "Online Privacy as Legal Safeguard: The Relationship Among Consumer, Online Portal, and Privacy Policies." *New Media and Society* 9 (5): 715–34.

Fernheimer, Janice W. 2016. "Confronting Kenneth Burke's Anti-Semitism." *Journal of Communication and Religion* 39 (2): 36–53.

Fisher, Walter R. 1994. "A Case: Public Moral Argument." In *The Reach of Dialogue: Confirmation, Voice, and Community*, edited by Anderson, Rob, Kenneth N. Cissna, and Ronald C. Arnett, 173–77. New York: Hampton Press.

Flower, Linda, and John R. Hayes. 1981. "A Cognitive Process Theory of Writing." *College Composition and Communication* 32 (4): 365–87.

Flyvbjerg, Bent. 2001. *Making Social Science Matter: Why Social Inquiry Fails and How It Can Succeed Again*. Cambridge: Cambridge University Press.

Flyvbjerg, Bent, Todd Landman, and Sanford Schram, eds. 2012. *Real Social Science: Applied Phronesis*. Cambridge: Cambridge University Press.

Foot, Philippa. 1978. *Virtues and Vices and Others Essays in Moral Philosophy*. Berkeley: University of California Press.

Freedman, David H. 2010. *Wrong: Why Experts Keep Failing Us—and How to Know When Not to Trust Them*. New York: Little, Brown.

Friedland, Martin L. 2002. *The University of Toronto: A History*. 2nd ed. Toronto: University of Toronto Press.

Gadamer, Hans-George. 1997. "Reflections on My Philosophical Journey." In *The Philosophy of Hans-George Gadamer*, edited by Lewis Edwin Hahn, 3–63. Chicago: Open Court.

Gage, John T. 2018. "What Is Rhetorical Phronesis? Can It Be Taught?" *Rhetoric Review* 37 (4): 327–34.

Gallagher, John R. 2018. "Enacting Virtue Ethics." *Rhetoric Review* 37 (4): 379–84.

———. 2019. *Update Culture and the Afterlife of Digital Writing*. Logan: Utah State University Press.

Garver, Eugene. 2017. "Aristotle's Rhetoric in Theory and Practice." In *The Oxford Handbook of Rhetorical Studies*, edited by Michael J. MacDonald, 133–41. Oxford: Oxford University Press.

Geertz, Clifford. 1983. *Local Knowledge: Further Essays in Interpretive Anthropology*. New York: Basic Books.

Gladkova, Olga, Chrysanne DiMarco, and Randy Allen Harris. 2015. "What's in a Name? Journal Titles in the Field of Epistemic Research." *Journal of Argumentation in Context* 3 (3): 259–86.

———. 2016. "Argumentative Meanings and Their Stylistic Configurations in Clinical Research Publications." *Argument and Computation* 6 (3): 310–46.

Gladwell, Malcolm. 2008. *Outliers: The Story of Success*. New York: Little, Brown.

Gobet, Fernand. 2011. *Psychologie du talent et de l'expertise*. Brussels: De Boeck.

———. 2016. *Understanding Expertise: A Multi-Disciplinary Approach*. London: Palgrave Macmillan.

Gobet, Fernand, Peter C. R. Lane, and Martyn Lloyd-Kelly. 2015. "Chunks, Schemata, and Retrieval Structures: Past and Current Computational Models." *Frontiers in Psychology* 6 (22), https://doi.org/10.3389/fpsyg.2015.01785.

Goodnight, G. Thomas. 1982. "The Personal, Technical, and Public Spheres of Argument: A Speculative Inquiry into the Art of Public Deliberation." *Journal of the American Forensic Association* 18 (4): 214–27.

Gould, Stephen Jay. 2003. *The Hedgehog, the Fox, and the Magister's Pox: Mending the Gap Between Science and the Humanities*. Three Rivers, CA: Three Rivers Press.

Graham, S. Scott, and Lynda Walsh [now Lynda Olman]. 2019. "There's No Such Thing as a Scientific Controversy." *Technical Communication Quarterly* 28 (3): 192–206.

Gross, Alan. G. 2006. *Starring the Text: The Place of Rhetoric in Science Studies*. Carbondale: Southern Illinois University Press.

Grundmann, Reiner. 2018. "The Rightful Place of Expertise." *Social Epistemology* 32 (6): 372–86.

Haldane, John. 2017. "Virtue Ethics in the Medieval Period." In *Varieties of Virtue Ethics*, edited by David Carr, James Arthur, and Kristján Kristjánsson, 73–88. London: Palgrave Macmillan.

Halloran, S. Michael. 1982. "Aristotle's Concept of Ethos, or If Not His Somebody Else's." *Rhetoric Review* 1 (1): 58–63.

Hambrick, David Z., Guillermo Campitelli, and Brooke N. Macnamara. 2017. "Introduction: A Brief History of the Science of Expertise and Overview of the Book." In *The Science of Expertise*, edited by David Z. Hambrick, Guillermo Campitelli, and Brooke N. Macnamara, 1–9. New York: Routledge.

———, eds. 2017. *The Science of Expertise*. New York: Routledge.

Hardwig, John. 1991. "The Role of Trust in Knowledge." *Journal of Philosophy* 88 (12): 693–708.

Hariman, Robert. 2003. "Prudence in the Twenty-First Century." In *Prudence: Classical Virtue, Postmodern Practice*, edited by Robert Hariman, 287–322. University Park: Penn State University Press.

Harper, Douglas. 1992. *Working Knowledge: Skill and Community in a Small Shop*. Berkeley: University of California Press. First published 1987.

Harris, Randy Allen. 2013a. "Figural Logic in Gregor Mendel's 'Experiments on Plant Hybrids.'" *Philosophy and Rhetoric* 46 (4): 570–602.

———. 2013b. "The Rhetoric of Science Meets the Science of Rhetoric." *POROI: The Project on Rhetoric of Inquiry* 9 (1), https://doi.org/10.13008/2151-2957.1158.

———, ed. 2018. *Landmark Essays on the Rhetoric of Science: Case Studies*. 2nd ed. New York: Routledge.

———, ed. 2019. *Landmark Essays on the Rhetoric of Science: Issues and Methods*. New York: Routledge.

Harris, Randy Allen, and Chrysanne Di Marco. 2009. "Constructing a Rhetorical Figuration Ontology." In *Persuasive Technology and Digital Behaviour Intervention Symposium*, edited by Judith Masthoff and Floriana Grasso, 47–52. Edinburgh: Society for the Study of Artificial Intelligence and Simulation of Behaviour (AISB).

Harris, Randy Allen, Chrysanne Di Marco, Ashley Rose Mehlenbacher, Robert Clapperton, Insun Choi, Isabel Li, Sebastian Ruan, and Cliff O'Reilly. 2017. "A Cognitive Ontology of Rhetorical Figures." In *Society for the Study of Artificial Intelligence and Simulation of Behaviour's (AISB) Cognition and Ontologies Workshop*, 224–31. Bath: Society for the Study of Artificial Intelligence and Simulation of Behaviour (AISB).

Hartelius, E. Johanna. 2011. *The Rhetoric of Expertise*. Lanham, MD: Lexington Books.

———. 2020. *The Gifting Logos: Expertise in the Digital Commons*. Berkeley: University of California Press.

Hayes, John R., and Linda Flower. 1981. "Uncovering Cognitive Processes in Writing: An Introduction to Protocol Analysis." In *Research on Writing*, edited by Peter Mosenthal, Lynne Tamar, and Sean A. Walmsley, 206–20. New York: Longman.

———. 1986. "Writing Research and the Writer." *American Psychologist* 41 (10): 1106–13.

Hayes, John R., Linda Flower, Karen A. Schriver, James Stratman, and Linda Carey. 1987. "Cognitive Processes in Revision." In *Advances in Applied Psycholinguistics*, edited by Sheldon Rosenberg, 176–240. Cambridge: Cambridge University Press.

Herndl, Carl G. 2016. "Doing and Knowing in the L'Aquila Case." *Social Epistemology Review and Reply Collective* 5 (6): 1–6.

Hofstadter, Richard. 1963. *Anti-Intellectualism in American Life*. New York: Vintage.

Hogarth, Robin M., Tomás Lejarraga, and Emre Soyer. 2015. "The Two Settings of Kind and Wicked Learning Environments." *Current Directions in Psychological Science* 24 (5): 379–85.

Horn, Michiel. 1999. *Academic Freedom in Canada: A History*. Toronto: University of Toronto Press.

Hursthouse, Rosalind. 2006. "Practical Wisdom: A Mundane Account." In *Proceedings of the Aristotelian Society* 106 (1): 285–309.

International Chess Federation. 2020. "FIDE Code of Ethics." In *FIDE Handbook*. Updated March 1, 2020. https://handbook.fide.com.

Isocrates. 1961. *Antidosis*. In *Isocrates*. Vol. 2. Translated by George Norlin. Loeb Classical Library. Cambridge, MA: Harvard University Press.

Jasanoff, Sheila. 1998. "Expert Games in Silicone Gel Breast Implant Litigation." In *Science in Court*, edited by Michael Freeman and Helen Reece, 83–107. Brookfield, VT: Ashgate.

———. 2003. "Breaking the Waves in Science Studies: Comment on H. M. Collins and Robert Evans, 'The Third Wave of Science Studies.'" *Social Studies of Science* 33 (3): 389–400.

Kakutani, Michiko. 2018. *The Death of Truth*. London: William Collins.

Katz, Steven B. 1992. "The Ethic of Expediency: Classical Rhetoric, Technology, and the Holocaust." *College English* 54 (3): 255–75.

———. 1993. "Aristotle's Rhetoric, Hitler's Program, and the Ideological Problem of Praxis, Power, and Professional Discourse." *Journal of Business and Technical Communication* 7 (1): 37–62.

Keith, William, and Robert Danisch. 2020. *Beyond Civility: The Competing Obligations of Citizenship*. University Park: Penn State University Press.

Kelly, Ashley Rose [now Ashley Rose Mehlenbacher], and Kate Maddalena. 2015. "Harnessing Agency for Efficacy: 'Foldit' and Citizen Science." *POROI—The Project on Rhetoric of Inquiry* 11 (1): 1–20.

———. 2016. "Networks, Genres, and Complex Wholes: Citizen Science and How We Act Together Through Typified Text." *Canadian Journal of Communication* 41 (2): 287–303.

Khan, Carrie-Ann Biondi. 2005. "Aristotle's Moral Expert: The Phronimos." In *Ethics Expertise: History, Contemporary Perspectives, and Applications*, edited by Lisa Rasmussen, 39–53. Dordrecht: Springer Netherlands.

Kimura, Aya Hirata. 2016. *Radiation Brain Moms and Citizen Scientists: The Gender Politics of Food Contamination After Fukushima*. Durham, NC: Duke University Press.

Kimura, Aya Hirata, and Abby Kinchy. 2016. "Citizen Science: Probing the Virtues and Contexts of Participatory Research." *Engaging Science, Technology, and Society* 2:331–61.

———. 2019. *Science by the People: Participation, Power, and the Politics of Environmental Knowledge*. New Brunswick, NJ: Rutgers University Press.

Kinsella, Elizabeth Anne. 2010. "The Art of Reflective Practice in Health and Social Care: Reflections on The Legacy of Donald Schön." *Reflective Practice* 11 (4): 565–75.

Kinsella, Elizabeth Anne, and Allan Pitman, eds. 2012. *Phronesis as Professional Knowledge: Practical Wisdom in the Professions*. Rotterdam: Sense Publishers.

Kinsella, William J. 2001. "Nuclear Boundaries: Material and Discursive Containment at the Hanford Nuclear Reservation." *Science as Culture* 10 (2): 163–94.

———. 2004. "Public Expertise: A Foundation for Citizen Participation in Energy and Environmental Decisions." In *Communication and Public Participation in Environmental Decision Marking*, edited by Stephen P. Depoe, John W. Delicath, and Marie-France Aepli Elsenbeer, 83–95. Albany: State University of New York Press.

Kinsella, William J., and Jay Mullen. 2007. "Becoming Hanford Downwinders: Producing Community and Challenging Discursive Containment." In *Nuclear Legacies: Communication,*

Controversy, and the U.S. Nuclear Weapons Complex, edited by Bryan C. Taylor, William J. Kinsella, Stephen P. Depoe, and Maribeth S. Metzler, 73–107. Lanham, MD: Lexington Books.

Knowles, Scott Gabriel. 2011. *The Disaster Experts: Mastering Risk in Modern America.* Philadelphia: University of Pennsylvania Press.

König, Jason. 2017. "Introduction: Self-Assertion and Its Alternatives in Ancient Scientific and Technical Writing." In *Authority and Expertise in Ancient Scientific Culture,* edited by Jason König and Greg Woolf, 1–26. Cambridge: Cambridge University Press.

Kuhn, Thomas S. 1970. *The Structure of Scientific Revolutions.* Chicago: University of Chicago Press. First published 1962.

Laski, Harold Joseph. 1931. *The Limitations of the Expert.* Fabian Tracts 235. London: Fabian Society.

Lave, Jean, and Etienne Wenger. 1991. *Situated Learning: Legitimate Peripheral Participation.* Cambridge: Cambridge University Press.

Levitan, Dave. 2017. *Not a Scientist: How Politicians Mistake, Misrepresent, and Utterly Mangle Science.* New York: W. W. Norton.

Lieberman, Jethro Koller. 1970. *The Tyranny of the Experts: How Professionals Are Closing the Open Society.* New York: Walker.

Lyne, John, and Henry F. Howe. 1990. "The Rhetoric of Expertise: E. O. Wilson and Sociobiology." *Quarterly Journal of Speech* 76 (2): 134–51.

Lynn, Kenneth S. 1963. "Introduction to the Issue 'The Professions.'" *Daedalus* 92 (4): 649–54.

MacIntyre, Alasdair. 1981. *After Virtue: A Study in Moral Theory.* London: Duckworth.

Macnamara, Brooke N., David Z. Hambrick, David J. Frank, Michael J. King, Alexander P. Burgoyne, and Elizabeth Meinz. 2017. "The Deliberate Practice View: An Evaluation of Definitions, Claims, and Empirical Evidence." In *The Science of Expertise,* edited by David Z. Hambrick, Guillermo Campitelli, and Brooke N. Macnamara, 151–68. New York: Routledge.

Mailloux, Steven. 2004. "Rhetorical Hermeneutics Still Again." In *A Companion to Rhetoric and Rhetorical Criticism,* edited by Wendy Jost and Walter Olmsted, 457–72. Malden, MA: Blackwell.

Maimonides. 1956. *The Guide of the Perplexed.* Translated by Michael Friedländer. New York: Dover Publications.

Majdik, Zoltan P. 2016. "Expertise as Practice: A Response to Devasto." *Social Epistemology Review and Reply Collective* 5 (11): 1–6.

Majdik, Zoltan P., and William M. Keith. 2011a. "Expertise as Argument: Authority, Democracy, and Problem-Solving." *Argumentation* 25 (3): 371–84.

———. 2011b. "The Problem of Pluralistic Expertise: A Wittgensteinian Approach to the Rhetorical Basis of Expertise." *Social Epistemology* 25 (3): 275–90.

Mehlenbacher, Ashley Rose. 2017. "Rhetorical Figures as Argument Schemes: The Proleptic Suite." *Argument and Computation* 8 (3): 233–52.

———. 2019. *Science Communication Online: Engaging Experts and Publics on the Internet.* Columbus: Ohio State University Press.

Mehlenbacher, Ashley Rose, and Randy Allen Harris. 2017. "A Figurative Mind: Gertrude Buck's *The Metaphor* as a Nexus in Cognitive Metaphor Theory." *Rhetorica* 35 (1): 75–109.

Mehlenbacher, Ashley Rose, and Carolyn R. Miller. 2018. "Intersections: Scientific and Parascientific Communication on the Internet." In *Landmark Essays on Rhetoric of Science: Case Studies,* edited by Randy Allen Harris, 239–60. 2nd ed. New York: Routledge.

Mehlenbacher, Brad. 2009. "Multidisciplinarity and 21st Century Communication Design." In *SIGDOC '09: The 27th ACM International Conference on Design of Communication Proceedings*, 59–65. Bloomington: ACM.

———. 2013. "What Is the Future of Technical Communication?" In *Solving Problems in Technical Communication*, edited by Johndan Johnson-Eilola and Stuart A. Selber, 187–208. Chicago: University of Chicago Press.

Mehta, Aalok, Zoltan P. Majdik, and Carrie Anne Platt. 2012. "Controversy, Conflict, and Conflicting Expertise: Report from the 2011 ARST Pre-Conference at NCA." *POROI: The Project on Rhetoric of Inquiry* 8 (1), http://www.doi.org/10.13008/2151-2957.1115.

Mercieca, Jennifer. 2020. *Demagogue for President: The Rhetorical Genius of Donald Trump*. College Station: Texas A&M University Press.

Metzger, David. 2014. "Maimonides's Contribution to a Theory of Self-Persuasion." In *Jewish Rhetorics: History, Theory, Practice*, edited by Michael Bernard-Donals and Janice W. Fernheimer, 112–30. Waltham, MA: Brandeis University Press.

Miller, Carolyn R. 1984. "Genre as Social Action." *Quarterly Journal of Speech* 70 (2): 151–76.

———. 1992. "Kairos in the Rhetoric of Science." In *A Rhetoric of Doing: Essays on Written Discourse in Honor of James L. Kinneavy*, edited by Stephen Paul Witte, Neil Nakadate, and Roger Dennis Cherry, 310–27. Carbondale: Southern Illinois University Press.

———. 2003. "The Presumptions of Expertise: The Role of Ethos in Risk Analysis." *Configurations* 11 (2): 163–202.

Miller, Christian B. 2018. *The Character Gap: How Good Are We?* New York: Oxford University Press.

Miller, George A. 1994. "The Magical Number Seven, Plus or Minus Two: Some Limits on Our Capacity for Processing Information." *Psychological Review* 101 (2): 343–52. First published 1956.

Millgram, Elijah. 2015. *The Great Endarkenment: Philosophy for an Age of Hyperspecialization*. Oxford: Oxford University Press.

Mol, Annemarie. 2002. *The Body Multiple: Ontology in Medical Practice*. Durham, NC: Duke University Press.

Moriarty, Devon, Paula Núñez De Villavicencio, Lillian A. Black, Monica Bustos, Helen Cai, Brad Mehlenbacher, and Ashley Rose Mehlenbacher. 2019. "Durable Research, Portable Findings: Rhetorical Methods in Case Study Research." *Technical Communication Quarterly* 28 (2): 124–36.

Motta, Matt. 2017. "Republicans Are Increasingly Antagonistic Toward Experts. Here's Why That Matters." *Washington Post*, August 11, 2017. https://www.washingtonpost.com /news/monkey-cage/wp/2017/08/11/republicans-are-increasingly-antagonistic-toward -experts-heres-why-that-matters.

Nichols, Tom. 2017. *The Death of Expertise: The Campaign Against Established Knowledge and Why It Matters*. New York: Oxford University Press.

Noble, Safiya U. 2018. *Algorithms of Oppression: How Search Engines Reinforce Racism*. New York: New York University Press.

Nussbaum, Martha. 1986. *The Fragility of Goodness: Luck and Ethics in Greek Tragedy and Philosophy*. Cambridge: Cambridge University Press.

———. 1999. "Virtue Ethics: A Misleading Category?" *Journal of Ethics* 3 (3): 163–201.

Ochigame, Rodrigo. 2019. "The Invention of 'Ethical AI': How Big Tech Manipulates Academia to Avoid Regulation." *The Intercept*, December 20, 2019. https://theintercept.com /2019/12/20/mit-ethical-ai-artificial-intelligence.

Ohanian, Roobina. 1990. "Construction and Validation of a Scale to Measure Celebrity Endorsers' Perceived Expertise, Trustworthiness, and Attractiveness." *Journal of Advertising* 19 (3): 39–52.

Ottinger, Gwen. 2010. "Buckets of Resistance: Standards and the Effectiveness of Citizen Science." *Science, Technology, and Human Values* 35 (2): 244–70.

———. 2013. *Refining Expertise: How Responsible Engineers Subvert Environmental Justice Challenges.* New York: New York University Press.

Pew Research Center. 2017. "Sharp Partisan Divisions in Views of National Institutions." Pew Research Center, July 10, 2017. https://www.people-press.org/2017/07/10/sharp -partisan-divisions-in-views-of-national-institutions.

———. 2019. "Trust and Mistrust in Americans' Views of Scientific Experts." Pew Research Center, August 2, 2019. https://www.pewresearch.org/science/wp-content/uploads /sites/16/2019/08/PS_08.02.19_trust.in_.scientists_FULLREPORT_8.5.19.pdf.

Pezzullo, Phaedra C. 2001. "Performing Critical Interruptions: Stories, Rhetorical Invention, and the Environmental Justice Movement." *Western Journal of Communication* 65 (1): 1–25.

Pfister, Damien Smith. 2011. "Networked Expertise in the Era of Many-to-Many Communication: On Wikipedia and Invention." *Social Epistemology* 25 (3): 217–31.

Pietrucci, Pamela, and Leah Ceccarelli. 2019. "Scientist Citizens: Rhetoric and Responsibility in L'Aquila." *Rhetoric and Public Affairs* 22 (1): 95–128.

Plaisance, Kathryn S., Alexander V. Graham, John McLevey, and Jay Michaud. 2019. "Show Me the Numbers: A Quantitative Portrait of the Attitudes, Experiences, and Values of Philosophers of Science Regarding Broadly Engaged Work." *Synthese* 198: 4603–33.

Plato. 1974. *The Republic.* Translated by Desmond Lee. Harmondsworth: Penguin. First printed 1955.

Polanyi, Michael. 1966. *The Tacit Dimension.* Chicago: University of Chicago Press.

Poulakos, John. 1984. "Rhetoric, the Sophists, and the Possible." *Communication Monographs* 51 (3): 215–26.

Poulakos, John, and Takis Poulakos. 1997. *Speaking for the Polis: Isocrates' Rhetorical Education.* Columbia: University of South Carolina Press.

Quast, Christian. 2018. "Towards a Balanced Account of Expertise." *Social Epistemology* 32 (6): 397–419.

Rice, Jenny. 2015. "Para-Expertise, Tacit Knowledge, and Writing Problems." *College English* 78 (2): 117–38.

Rip, Arie. 2003. "Constructing Expertise: In a Third Wave of Science Studies?" *Social Studies of Science* 33 (3): 419–34.

Robson, David. 2019. *The Intelligence Trap: Why Smart People Do Stupid Things and How to Make Wiser Decisions.* London: Hodder & Stoughton.

Roundtree, Aimee K. 2020. "ANT Ethics in Professional Communication: An Integrative Review." *American Communication Journal* 22 (1): 1–13.

Russell, Sharman Apt. 2014. *Diary of a Citizen Scientist: Chasing Tiger Beetles and Other New Ways of Engaging the World.* Eugene: Oregon State University Press.

Sayers, Dorothy L. 2005. "Are Women Human? Address Given to a Women's Society, 1938." *Logos: A Journal of Catholic Thought and Culture* 8 (4): 165–78.

Schlagwein, Felix. 2020. "The Real 'Queen's Gambit': Judit Polgar." Translated by Sarah Hucal. *Deutsche Welle* (Bonn and Berlin), December 9, 2020. https://p.dw.com/p/3mSXn.

Schön, Donald. 1983. *Reflective Practitioner: How Professionals Think in Action.* New York: Basic Books.

Schwartz, Barry, and Kenneth Sharpe. 2010. *Practical Wisdom: The Right Way to Do the Right Thing.* New York: Riverhead Books.

Self, Lois S. 1979. "Rhetoric and Phronesis: The Aristotelian Ideal." *Philosophy and Rhetoric* 12 (2): 130–45.

Seneca. 1965. *Ad Lucilium Epistulae Morales*. In *L. Annaei Senecae*. Vol. 1. Edited by L. D. Reynolds. Oxford Classical Texts. Oxford: Clarendon Press.

Sherman, Nancy. 1991. *The Fabric of Character: Aristotle's Theory of Virtue*. Oxford: Clarendon Press. First published 1989.

Sim, May. 2007. *Remastering Morals with Aristotle and Confucius*. Cambridge: Cambridge University Press.

———. 2018. "The Phronimos and the Sage." In *The Oxford Handbook of Virtue*, edited by Nancy E. Snow, 190–205. Oxford: Oxford University Press.

Sismondo, Sergio. 2017. "Post-Truth?" *Social Studies of Science* 47 (1): 3–6.

Skloot, Rebecca. 2010. *The Immortal Life of Henrietta Lacks*. New York: Broadway.

Sloane, Thomas O. 2001. *Encyclopedia of Rhetoric*. Oxford: Oxford University Press.

Snow, Charles Percy. 2012. *The Two Cultures*. Cambridge: Cambridge University Press. First published 1959.

Solomon, Martha. 1985. "The Rhetoric of Dehumanization: An Analysis of Medical Reports of the Tuskegee Syphilis Project." *Western Journal of Speech Communication* 49 (4): 233–47.

Sorabji, Richard. 1972. *Aristotle on Memory*. Providence: Brown University Press.

Sternberg, Robert J. 2017. "Four Kinds of Expertise." In *The Science of Expertise*, edited by David Z. Hambrick, Guillermo Campitelli, and Brooke N. Macnamara, 419–26. New York: Routledge.

Sutton, Robert. 2010. *The No Asshole Rule: Building a Civilized Workplace and Surviving One That Isn't*. New York: Grand Central Publishing.

Swales, John M. 1990. *Genre Analysis: English in Academic and Research Settings*. Cambridge: Cambridge University Press.

———. 2004. *Research Genres: Explorations and Applications*. Cambridge Applied Linguistics Series. Cambridge: Cambridge University Press.

Tindale, C. W. 2012. "Dismantling Expertise: Disproof, Retraction, and the Persistence of Belief." In *Between Scientists and Citizens: Proceedings of a Conference at Iowa State University, June 1–2, 2012*, edited by Jean Goodwin, 393–402. Ames, IA: Great Plains Society for the Study of Argumentation.

Todd, Sarah. 2019. "Finally, a Performance Review Designed to Weed out 'Brilliant Jerks.'" *Quartz*, July 22, 2019. https://qz.com/work/1671163/atlassians-new-performance-review-categories-weed-out-brilliant-jerks.

Toulmin, Stephen Edelston. 2001. *Return to Reason*. Cambridge, MA: Harvard University Press.

Useem, Jerry. 2019. "At Work, Expertise Is Falling Out of Favor." *The Atlantic*, June 22, 2019. https://www.theatlantic.com/magazine/archive/2019/07/future-of-work-expertise-navy/590647.

Vettese, Troy. 2019. "Sexism in the Academy: Women's Narrowing Path to Tenure." *n+1* 34 (Spring), https://nplusonemag.com/issue-34/essays/sexism-in-the-academy.

Vivian, Bradford. 2018. "Memory: *Ars Memoriae*, Collective Memory, and the Fortunes of Rhetoric." *Rhetoric Society Quarterly* 48 (3): 287–96.

Waddell, Craig. 1990. "The Role of Pathos in the Decision-Making Process: A Study in the Rhetoric of Science Policy." *Quarterly Journal of Speech* 76 (4): 381–400.

Wai, Jonathan. 2014. "Experts Are Born, Then Made: Combining Prospective and Retrospective Longitudinal Data Shows That Cognitive Ability Matters." *Intelligence* 45:74–80.

Wai, Jonathan, and Harrison J. Kell. 2017. "How Important Is Intelligence in the Development of Professional Expertise? Combining Prospective and Retrospective Longitudinal

Data Provides an Answer." In *The Science of Expertise*, edited by David Z. Hambrick, Guillermo Campitelli, and Brooke N. Macnamara, 73–86. New York: Routledge.

Walker, Kenneth, and Lynda Walsh [now Lynda Olman]. 2012. "'No One Yet Knows What the Ultimate Consequences May Be': How Rachel Carson Transformed Scientific Uncertainty into a Site for Public Participation in Silent Spring." *Journal of Business and Technical Communication* 26 (1): 3–34.

Walwema, Josephine. 2020. "The WHO Health Alert: Communicating a Global Pandemic with WhatsApp." *Journal of Business and Technical Communication* 35 (1): 35–40.

Walton, Douglas. 1992. *The Place of Emotion in Argument*. University Park: Penn State University Press.

Wasserman, Mira Beth. 2020. "Against Apocalyptic Ethics: Human Responsibility Before, During, and After a Pandemic." *Evolve*, August 17, 2020. http://evolve.reconstructing judaism.org/againstapocalypticethics.

Weinberg, Alvin M. 1972. "Science and Trans-Science." *Minerva* 10 (2): 209–22.

———. 1992. *Nuclear Reactions: Science and Trans-Science*. New York: American Institute of Physics.

White, Morton. 1962. "Reflections on Anti-Intellectualism." *Daedalus* 91 (3): 457–68.

Whyte, Kyle Powys, and Robert P. Crease. 2010. "Trust, Expertise, and the Philosophy of Science." *Synthese* 177 (3): 411–25.

Williams, Joan C., and Marina Multhaup. 2018. "For Women and Minorities to Get Ahead, Managers Must Assign Work Fairly." *Harvard Business Review*, March 5, 2018. https://hbr.org/2018/03/for-women-and-minorities-to-get-ahead-managers-must -assign-work-fairly.

Wilson, Edward O. 1971. *The Insect Societies*. Cambridge, MA: Harvard University Press.

———. 1975. *Sociobiology: The New Synthesis*. Cambridge, MA: Harvard University Press.

———. 1978. *On Human Nature*. Cambridge, MA: Harvard University Press.

Wittgenstein, Ludwig. 2009. *Philosophical Investigations*. Translated by G. E. M. Anscombe, P. M. S. Hacker, and Joachim Schulte. Hoboken, NJ: John Wiley & Sons. First published 1953.

Wynne, Brian. 1989. "Sheepfarming After Chernobyl: A Case Study in Communicating Scientific Information." *Environment: Science and Policy for Sustainable Development* 31 (2): 10–39.

———. 1992. "Misunderstood Misunderstanding: Social Identities and Public Uptake of Science." *Public Understanding of Science* 1 (3): 281–304.

Xiong-Gum, Mai Nou. 2018. "Place as Interface, Sensory-Data, and Phronesis." In *Proceedings of the Annual Computers and Writing Conference*, edited by Chen Chen, Kristopher Purzycki, and Lydia Wilkes, 65–74. Fort Collins, CO: WAC Clearinghouse.

Yates, Frances A. 1966. *The Art of Memory*. Chicago: University of Chicago Press.

Yuan, Ying. 2017. "The Argumentative Litotes in *The Analects*." *Argument and Computation* 8 (3): 253–66.

Zagzebski, Linda Trinkaus. 1996. *Virtues of the Mind: An Inquiry into the Nature of Virtue and the Ethical Foundations of Knowledge*. Cambridge: Cambridge University Press.

———. 2017. *Exemplarist Moral Theory*. Oxford: Oxford University Press.

Index